AROUND
THE WORLD
IN

200
GLOBES

Stories of the Twentieth Century

Willem Jan Neutelings

LUSTER

A Blue Marble in a Black Universe

On Christmas Eve 1968, for the first time in history, a manned spaceship is about to fly around the Moon. Millions of people on Earth are glued to their TV sets. To mark the drama of the moment, the astronauts on board the Apollo 8, Frank Borman, James Lovell, and William Anders, read aloud the first ten verses of Genesis: The Creation of the Earth. When their spaceship turns around the Moon and redirects itself towards the Earth, the three men are stunned: they are the first people to see the Earth as a splendid blue globe, floating in an infinite black universe. On December 24, 1968, the image of the Earth was finally recorded.

For more than five hundred years, globe makers have tried to construct an accurate image of the Earth, without ever being able to see it. By the beginning of the twentieth century, they succeeded in filling the last blind spots on the globe. God's natural world, with its rivers and coastlines, its mountains and deserts, found its correct shape and location on the scale model of the Earth. Man's imaginary world, however, with its political borders and shipping lines, its latitudes and time zones, is not visible from the Moon. The challenge for twentieth-century globe makers was to absorb the rapid developments in society and to find ways of representing them on a globe. Consequently, the design and production of globes altered dramatically. The evolution of globes followed the rhythm of inventions and innovations, such as plastics and computers. Globe designs adapted to architectural fashions, ranging from art nouveau to modernism. Globe cartography was revised to portray political and economic upheavals, and the globe makers themselves transformed from traditional workshops into industrial enterprises, amidst wars and crises.

This book is a chronicle of the turbulent twentieth century, told through the many extraordinary globes that were created during that era. Throughout these hundred years, the globe evolved from a prestigious object for the few, to a mass-produced item accessible to all. Along its journey, the popular scale model of the Earth transformed into a teaching tool for schools and an interior decoration for the middle class. Its final stage is the generic plastic globe, comforting companion of children, radiating its gentle glow in millions of bedrooms around the world.

Willem Jan Neutelings

1
Earth, the Mother of all globes.
Earthrise, photo taken by Apollo 8 astronaut
William Anders on Christmas Eve 1968

INTRODUCTION

A globe is a mirror of its time. Wether a paper globe, handmade by Gerard Mercator in the sixteenth century, or a plastic globe, made in the twentieth century by a machine, each tells a unique story that reflects the state of technology, economy, politics, art, and science at the moment of its creation.

Globe scholars often draw a sharp line between the traditional handmade globes produced until the mid-nineteenth century, and the modernist mass-produced globes of the twentieth century. However, the journey of the globe is an uninterrupted narrative of human history, a history that continued throughout the twentieth century, up to the present day. Plenty of books and articles can be found on antique globes, but little research exists on the globes of the past hundred years. One reason may be that the scarce antique globes are sold at classy auctions for the price of a sports car, while twentieth-century globes exist in vast numbers and are sold on eBay for the price of a parking ticket.

Meanwhile, interest in twentieth-century globes surges in the slipstream of vintage mid-century modern design, the highly sought-after modernist furniture. After more than two decades living in the twenty-first century, we can take sufficient distance and look back at these vintage globes, which reflect so well the turbulent times of seventy years ago. This book is a chronicle of the twentieth century and its radical changes, eloquently told by these wondrous round objects.

In 1914, at the outbreak of the First World War, the nineteenth century finally came to an end. This cruel event shattered the joyful bourgeois lifestyle of the Belle Epoque. Empires were broken up, and a new world order emerged. It was a time of political revolutions, social changes, and disruptive technologies, such as the car, the airplane, and electricity. The middle class rose to dominance, and the possession of a globe, once a scarce object for the privileged few, became aspirational for a broad section of the population. The globe makers of the 1920s redesigned the globes to make them fit for middle-class interiors. *"A globe in every house!"* was the slogan in Europe as well as in America. The young and ambitious globe makers radically chose the future, by developing new methods to mass-produce globes in high numbers, at great speed, and at low costs, while pushing traditional globe makers out of the market.

In the same period, arts and architecture broke with the historical styles of the past. New styles such as art nouveau, art deco, and modernism emerged and rapidly succeeded each other. The role of globes transformed from scientific instruments to decorative objects, and globe makers swiftly adapted their designs to align with the latest fashion trends. To meet contemporary tastes, they redesigned the cartography, colored the oceans silver or black, and streamlined the metal stands, all to please the eye of the client.

The interwar period was a time of heroic explorations, such as the first flights by airplane across the Atlantic or Zeppelins above the North Pole. The exceptional events, closely followed on the radio, the new communication medium, fueled the excitement of the general public. Globe makers kept up with the heartbeat of the time and published new models with adapted cartography to incorporate these endeavors. In those years, globes were densely packed with information. Every aspect of society and economy was represented, from radio-stations, shipping lines, air-routes to even caravan trails, all depicted with lines and icons.

Another factor that transformed globes during the twentieth century was the progress in science and education. Radical laws aimed at combatting illiteracy required children to attend school longer and promoted new pedagogical methods. Geography became a compulsory subject in the school curriculum, and globes became a mandatory fixture in

Soviet workers assembling the globe that is integrated in the facade of the Central Telegraph building /
Moscow, 1928

INTRODUCTION

classrooms. As a result, school supply companies monopolized the market for educational globes in Europe and the U.S. They developed special models, such as inexpensive small globes for every child's desk, suspended globes to hang from classroom ceilings, cradle globes that could be picked up by students, unbreakable globes designed to withstand being dropped on the floor, and more. The educational globe makers also simplified the cartography, adapting it to the cognitive levels of school children. Models were offered with various levels of information to address different school grades.

Geographical sciences made significant progress in the twentieth century, and new fields of research emerged. One way to disseminate this new knowledge was through the creation of thematic globes, each demonstrating a specific aspect of Earth sciences, such as meteorology, geology, oceanography, and more. Twentieth-century globe makers constantly faced the challenge of balancing political globes, which depicted the world of man, with physical globes, which depicted the world of nature. The problem was solved in the 1950s with the invention of the dual-map globe, which alternatively shows the political or the physical map by switching the embedded illumination on or off.

During the twentieth century, permanent political changes were a constant headache for globe makers. The First and Second World Wars, the decolonization of Africa and Asia, the collapse of the U.S.S.R., and the breakup of Yugoslavia, all necessitated continual redrawing of political maps. Globes risked becoming quickly outdated, leaving globe makers with unsellable stock. On the brink of bankruptcy, some kept their sales afloat by accompanying globes with vouchers for postwar revision.

After the devastations of the Second World War in Europe, the situation on both sides of the Atlantic was very different. Large globe factories in Germany and Italy were destroyed by Allied bombings and had to be rebuilt. When the iron curtain fell, a sharp line was drawn between the Soviet block and the West-

ern block. Several German globe-making families fled to the West and had to restart their companies, while those that remained in the East saw their companies nationalized by the communist state.

Meanwhile, American globe makers continued to develop new models. American troops were engaged in conflicts all over the globe, while commercial aviation simultaneously experienced a significant boom. This triggered a paradigm shift in American politics and geography. The traditional concept of an isolated U.S. in the center of the map, protected by two oceans, was abandoned in favor of the Air-Age concept, which emphasized proximity to the other continents by using the North Pole as the center of the map. A new range of globes was created, which included measuring instruments for flight time and distance. Cradle stands allowed air travelers to pick up the globe and observe the new boundaryless Air-Age world from all angles.

In the aftermath of the Second World War, many babies were born during the so-called baby boom. American globe makers responded to the new children's market by introducing geographical game globes, which combined education and play. Small tin toy globes had already existed before the war, but these larger metal models contained more information and elevated the concept of a children's globe to a serious level.

The popular fascination with globes, these small models of the planet on which we live, made them attractive objects for appropriation beyond their original geographical purpose. In the twentieth century, globes were cleverly disguised as radios or as clocks, hollowed out to serve as coin banks or pencil sharpeners, or even transformed into housebars filled with whiskey bottles. They were also used as canvases for advertising slogans promoting air travel or car tires, and were deployed as propaganda tools for religious organizations, including the Pope.

Much like their predecessors, twentieth-century globe makers produced large floor globes on demand alongside their line of standardized industrial globes. These monumental globes often adorned the offices of presidents or businessmen, serving as symbolic expressions of power. Fashion models and movie stars, a new phenomenon in the twentieth century, used glamorous globes for their promotion as public figures in an increasingly mediatized society. Large institutions and international corporations frequently placed monumental custom-made globes in the lobbies of their headquarters, symbolizing their global reach.

In 1957, the Space Race begun when the Soviets successfully launched Sputnik, the first satellite to orbit the Earth. During the Cold War, the two superpowers, the United States and the Soviet Union, competed to be the first to reach the Moon. American globe makers released satellite globes featuring a small revolving ball, while Soviet globe makers created the first Moon globe which included the far side of the Moon, never before seen by humans. The moonshot captured the imagination of the general public, who followed the adventures of astronauts and cosmonauts at home on their television sets. By the time the first men landed on the Moon in 1969, sales of lunar globes peaked. However, the Moon frenzy quickly subsided, and planetary globes became a niche product once again.

Innovation was the driving force of the twentieth-century globe industry. On both sides of the Atlantic, numerous patents were granted, and many new models were created. The electric illuminated globe emerged in the 1920s, followed by the cradle globe in the 1930s. The 1940s saw the introduction of the transparent globe, while the dual-map globe appeared in the 1950s. The inflatable globe became popular in the 1960s, and the 1970s brought the lunar globe. The 1980s introduced the magnetic levitation globe, and the digital globe was developed in the 1990s, along with many other innovative models.

New synthetic materials, such as Bakelite, nylon, vinyl, or Plexiglas, emerged in the first half of the century. These 'plastics', as they were commonly called, initially proved useful for the military during the Second World War. Afterwards, they flooded society as colorful messengers of a new era, in the form of cheap and lightweight household objects. In the second half of the twentieth century, the development of plastics intertwined with the transformation of globes. Its final form became a printed plastic sphere, mounted on a plastic base, mass-produced on a conveyor belt. Plastics successfully spread to every corner of the planet, but by the end of the century, the dream had turned into a nightmare.

Most nineteenth-century globe makers began as family businesses producing road maps, school charts, or atlases, later diversifying into globes. However, the twentieth century saw the emergence of many newcomers who started their globe companies from scratch. The successful ones expanded their workshops into large globe industries, exporting globes to numerous countries in various languages. The last decades of the century were marked by industrial globalization and economical scaling. Many globe companies merged, were sold to large publishing conglomerates or simply ceased operations. Only a handful of large globe producers survived into the next millennium.

At the end of the twentieth century, the rapid development of computers and satellite systems caused a significant shift in the cartographic industry. Computer-generated maps, assembled from digital geographical data, replaced hand-drawn printed maps. The amount of geographical information that can be stored on a digital globe today, such as the Google Earth application on an iPhone, is infinitely more than what any traditional globe can hold. By the turn of the millennium, the purpose of the globe had diminished to serving as a soft glowing decoration for children's bedrooms, while a small digital globe, abundant with geographical information, nested in everybody's pocket.

Gerard Mercator
would have loved it!

TABLE OF CONTENTS

TECH-
NOL-
OGY

A Woman's World

During the twentieth century, the role division between men and women in Europe as well as in the United States remained largely traditional. While all globe companies were founded and directed by men, the labor force consisted predominantly of women who manufactured the globes.

In the beginning of the century, many traditional globe-making techniques were still in use. These included forming papier-mâché hemispheres by hand, cutting cartography out of printed sheets with scissors, glueing the map gores on the hemispheres with starch, and making color retouches with a fine brush. This delicate manual work was considered best suited for women. Most globe companies expanded their workshops and mechanized the manufacturing process. They invested in heavy machinery to mass-produce globes, including cardboard steam presses for forming hemispheres, vacuum heaters for molding plastic globes, automatic spray cabins, or conveyor belts to transport globes throughout the process. Once again, the monotonous work was deemed best suited for women, leading to the predominance of female operators for these machines.

The photograph shows a woman in the Rand McNally globe factory in Chicago in the 1950s, operating a heavy press. She is seen putting cardboard sheets with printed maps in the machine, which then presses them into hemispheres.

Technology changed rapidly in the twentieth century, and the globe had to catch up. The traditional plaster and paper sphere was replaced by metal, glass or plastic. Inventions and innovations made globes inflatable, foldable and electric. Economic and social changes went hand in hand: globes evolved from expensive handcrafted objects for the upper class, to cheap industrial products for the masses.

PAPER
GLOBES

Assembly Line

Henry Ford (1863-1947) the American entrepreneur, inspired by the mechanized slaughterhouses of Chicago, introduced the assembly line in his Detroit car factory in the 1910s. He optimized the production process of a car by splitting it into separate actions. Instead of a group of mechanics constructing a car together at fixed point, the car would move on a conveyor belt from worker to worker, with each one adding a specific piece to the car in a sequential process. In his movie *Modern Times*, film maker Charles Chaplin (1889-1977) criticized the stifling monotony of this repetitive labor. In the twentieth century, production of globes followed a similar path: the traditional craftsman's globe, once handmade in limited series as a luxury item for the wealthy, evolved into an industrial mass-produced commodity for the general public, conceived on an assembly line.

Orange Peels

The basis for traditional globes was a papier-mâché sphere, which was formed in a wooden mold. To smooth its surface, several gypsum layers were applied. Meanwhile, cartographic maps were engraved on copper plates, printed on paper, and cut into twelve sections resembling orange peels, known as map gores. Women meticulously hand glued the map gores onto the spheres, stretching them to fit the curvature. It was a precision work: one had to avoid any discontinuity in the map at the seam. A varnish was applied to protect the paper, and then the finished sphere was mounted on a wooden stand. Some globes had a metal meridian attached at the poles, allowing them to spin. However, the serial hand manufacturing was a slow process, resulting in globes being expensive objects. Around 1900, a large workshop could produce ten thousand globes a year at most.

Flower Petals

Innovative globe makers invested in steam presses to produce cardboard hemispheres. The smooth machine-made surfaces made the time-consuming addition of a plaster layer obsolete and allowed for accelerated production. Map application was also mechanized: circular maps, printed on cardboard, were machine cut into flower-petal sections and pressed together, forming ready-made hemispheres with a map surface. Two hemispheres slide together to form a globe, which then moves further along on the conveyor belt to be spray coated and hard baked. In the postwar period, output numbers surged, and production costs diminished. An industrial globe factory today can produce up to one million globes a year. As a result, twentieth-century globes transformed from luxury handmade objects into affordable machine-made mass products.

Assembly Line
Interior view of Replogle's industrial globe production plant / *Chicago, 2000s*

2

The George F. Cram Company from Indianapolis produced this ten-inch globe around 1939. It can easily be recognized as an American globe by the protruding tape on the equator, a common method used to fix two ready-pressed, map-covered hemispheres together and obscure the seam. The globe is spinning in a full meridian made from metal with a grade indication and mounted on a metal pin that sets the axis of the globe to the inclination of the Earth. The base, a four-toed wooden stand, seems inspired by the traditional American 1930s swivel office chair.

COLOR

GLOBES

Four-Color Theorem

South-African mathematician Francis Guthrie (1831-1899), while trying to color a map of the counties of England in 1852, noticed that only four different colors were needed. His Four-color Theorem is a mathematical conjecture that claims that any political map can be drawn with only four colors, without two countries of the same color touching each other. It means that the number of colors on a globe can be limited to four, saving on the number of printing cycles. For visual clarity, however, globe makers generally applied a larger number of colors. The traditional coloring of globes was a labor-intensive work. Before the advent of color-printing techniques, maps were engraved and printed in black and white. Subsequently, color had to be applied manually, a painstaking precision task usually entrusted to female colorists. This slow manual process increased the price of a traditional globe.

Chromolithography

French-German printer Godefroy Engelmann (1788-1839) patented chromolithography in 1837, a technique for printing multi-colored images. Maps are drawn in wax on a flat limestone and washed with acid, so ink adheres only to the surfaces requiring color. In the U.S., a comparable technique was developed in which a wax drawing is made on a metal sheet and then electroplated to create a raised relief metal printing block. The workforce at German map maker Perthes illustrates the transition: in 1881 the company employed ninety female hand colorists, by 1935, only one was left. Later, offset photolithography was invented: a drawing is applied photographically onto a metal plate, the image is inked, and subsequently transferred to a rubber-covered cylinder that prints the image on paper. The multi-color printing of maps allowed mass production of globes.

Border or Area Coloring

Two different coloring methods exist to separate countries on a globe. In the nineteenth century, border coloring was the common method: thick lines of saturated colors are applied to the borders of countries, while the rest of their surfaces remain uncolored. The advantage is that topographical names read better on a light background, the disadvantage is that erratic border lines can create a messy composition. In the twentieth century, area coloring became the common method: the surface of each country is covered in a single color, turning the map into a vivid patchwork. Area coloring emphasizes the shape of each country. The idiosyncratic shapes feed national symbols ingrained in our collective memory: after a century of area coloring, we immediately recognize the Italian 'boot' or the French 'hexagon', or the Oklahoma 'panhandle'.

Finishing Touch
Employee hand spraying a varnish on globes / *George Philip & Son factory / U.K., 1944*

3

Taride, a prominent twentieth-century Parisian map and globe maker, created this colorful, art deco-styled, glass globe in the late 1940s. The small fifteen-centimeter illuminated glass sphere has no meridian but sits directly on a piece of black marble. Its paper map gores are printed with strong contrasts and saturated colors, based on area coloring. The large monocolored surfaces representing the various European powers in Africa highlight the ongoing colonization of the continent. The dazzling red and blue lines in the mint-blue ocean indicate the course of hot and cold currents.

CAST-IRON

British Bananas

English gardener Joseph Paxton (1803-1865) experimented with the cultivation of tropical plants. He created the Cavendish Banana, named after his employer, which would become the universal banana, replacing common varieties worldwide after the devastating Panama disease outbreak in the 1950s. To grow bananas in the British climate, Paxton developed greenhouses made of glass and cast iron. This experience led to his winning the architectural competition for the Crystal Palace, a monumental building for the 1851 London World Exhibition. It was constructed entirely of prefabricated cast-iron elements and glass plates, bolted together onsite. The Crystal Palace was built in eight months, at a fraction of the cost of a traditional building, propelling cast iron to its status as the iconic material of the nineteenth century.

Fin-de-Siècle

Objects of any form or shape can be made by casting iron into a sand mold. The cast iron industry published catalogs with thousands of products made of this versatile material, available for order, including lampposts, fences, furniture, or radiators. Globes with a cast-iron base were on the market from the 1870s until the 1920s, when they were largely supplanted by sheet metal and Bakelite stands. A traditional stand was handmade of turned wood, a complex piece of furniture. Cast-iron stands enabled serial production but still allowed for abundant decoration, such as plant leaves or other relief patterns in fin-de-siècle style. The sphere itself remained a traditional hollow ball of plaster or paper, as cast-iron balls would be heavy and impractical. An exception to this rule is a number of miniature bank models, which have a cast-iron sphere.

American Colonies

The Grey Iron Casting Company of Mount Joy, Pennsylvania, produced miniature bank globes around 1900. Despite their small size—only three inches wide—the catalog claimed: *"All of our banks have a slot large enough to accommodate a half dollar or English penny."* The monochrome red hemispheres can be unscrewed to collect the coins. Landmasses are cast in relief. The only geographical names mentioned in golden letters are Hawaii and Philippines, commemorating the colonization of both nations by the U.S. in 1898, a policy *"to bring civilization"* [no. 82]. In the same period, globe maker Rand McNally collaborated with the American Globe and School Supply in Seneca Falls on a series of three-inch paperweight globes. These wooden spheres were finished with paper map gores and featured uncommon stands, such as a chandelier-like cast-iron base, decorated with lotus leaves.

Crystal Palace
a structure of prefabricated cast-iron elements and glass plates /
London, 1851

4

Windels, a Brussels globe maker, created this thirty-two-centimeter globe with a cast-iron tripod stand in the early twentieth century. Once a fashionable model, its lush style was rather outdated by the third edition in the mid-1920s. The globe reflects the racial theories of the period, at a time when Congo was still a Belgian colony. America, Asia, and Africa, are represented each by drawings of human heads of the 'Red, Yellow or Black Race', which were supposed to correspond to the inhabitants of each continent. Notably, the White Race was not limited to one continent, as printed on the globe: *"dispersed everywhere, to civilize and conquer."*

TIN

GLOBES

Cans and Fridges

Electrical engineer Alfred W. Mellowes (1879-1960), revolution-ized food conservation when he invented the home refrigerator in Fort Wayne, Indiana. His company, Frigidaire, began selling elec-trical fridges in 1915. However, it would take another fifty years be-fore every American household owned a fridge. In the meantime, the tin can remained the best solution for preserving food. The tin food can was invented in the early nineteenth century, around the same time as German metalworkers created tin toys. Originally, tin toys were folded from metal sheets, hand painted, and sold as expensive objects. By the end of the nineteenth century, tin sheets would be machine folded and color lithographed, transforming tin toys into cheap and mass-produced items. It was around this time that tin globes made their appearance in the German tin toy industry [no. 56].

Practically Indestructible

Metal globes have been around since antiquity, long before the ar-rival of paper globes. Tin globes represent the industrial version: a map is color lithographed onto a flat metal sheet and pressed into a mold to form two hollow hemispheres. When German exports halted during the First World War, U.S. tin toy makers like Ohio Art and J. Chein jumped into the market. Due to their limited sur-face, older three-inch tin globes feature rudimentary cartography. From the 1920s onwards, larger models were launched, up to sev-en inches in diameter, boasting detailed map images [nos. 14, 120-122]. Tin globes were promoted as robust competitors to the frag-ile paper globes. School supplier Denoyer-Geppert, for example, promoted its fifty cents, four-inch All-Metal Pupils Globe: *"Where cardboard globes are short-lived ... this new globe is ALL-METAL ... the globe is practically indestructible."* [no. 58].

Safety Rules

The postwar baby boom ignited the children's globe market. Sev-eral European toy factories, such as the French Jouets Mont-Blanc, the German Michael Seidel or the English Chad Valley, started producing lithographed metal globes for children [nos. 12, 122, 123]. Globes measuring up to twenty-four centimeters in diameter al-lowed for detailed cartography comparable to paper globes. In the U.S., Replogle produced eight-inch metal game globes, com-bining education and play [nos. 131-133]. During the 1960s Space Race, toy makers created metal Moon globes [nos. 33, 134-136]. Tin globes faced headwinds when the U.S. obliged Japan to create low value-added products. As a result, cheap 'Made in Japan' tin toys flooded the market, followed by inexpensive plastic globes. In the 1970s, consumer safety rules dealt a final blow to tin globes by imposing a ban on sharp-edged toys.

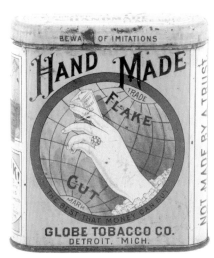

**Hand Made
Flake Cut**
Lithographed tin tobacco box /
Globe Tobacco Co. / Detroit, early 1900s

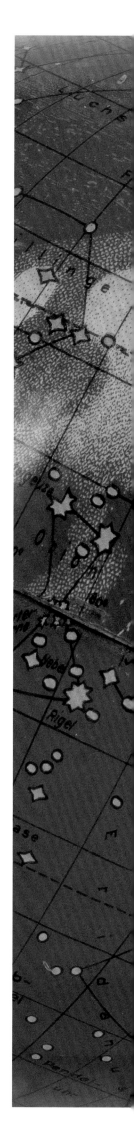

5
This small ten-centimeter celestial globe lacks the name of a company or author, but the tin stand and meridian resemble those of a 1930s model of the English brand Chad Valley. On the other hand, the lithographed map on the eleven-centimeter tin sphere is identical to the German-language map gores, created by astronomer Johannes Riem (1868-1943) and cartographer C. Luther for the celestial globes of the German globe maker Columbus Verlag [NO. 93]. It represents stars in magnitudes one to five, shows the constellations as line-connected stars, and depicts the galaxy as a white cloud.

ILLUMINATED GLASS

Electricity in the House

American inventor Thomas Edison (1847-1931) crafted the electric light bulb in 1879. It burned for thirteen hours in his laboratory in Menlo Park, New Jersey. Before that moment, artificial light was a dangerous luxury. Homes were dimly lit by flammable and expensive sources such as candles, oil, or gas. Electricity was considered a sign of progress, as illustrated by the famous 1920s slogan "*Communism = Soviets + Electricity*" coined by Vladimir Ilyich Lenin (1870-1924), the first leader of the Soviet Union. The electrification transition was an enormous undertaking. Not only did vast infrastructure for producing and distributing electricity need to be built, but also technical standards had to be regulated by law, including voltage levels. By 1930, approximately seventy percent of United States homes were equipped with electrical light.

Elegant Salons

Leipzig globe maker Paul Räth (1881-1929) invented the illuminated globe in 1920, by glueing paper map gores onto a glass sphere with an electric bulb inside. The cable was concealed in the wooden stand, and the switch was integrated into its base. Berlin competitor Columbus Verlag and many other globe makers soon followed with similar models. Following the fashion of female bronzes carrying an illuminated sphere, illuminated globes were promoted as elegant objects for the living room. In *Le Monde Illustré*, Parisian globe maker Girard & Barrère exclaimed: "*Our creation The Electric Illuminated Globe, the modern lamp par excellence, is the great novelty of the day. It conquered the most elegant salons, where it puts a discreet note of bluish light on the bronze art pieces, as well as in the office of the businessman, happy to pursue his plans to conquer a world where light shines clear!*" [no. 35].

Illuminated Plastic

During the 1950s, many globe makers sold illuminated globes as interior decoration. These sophisticated radiating spheres floated on elegant metal stands [nos. 139, 147, 149, 152]. The French brand Taride even had electric plugs made with their company logo cast into them to fit the globes. To reduce costs, the fragile glass sphere was soon replaced by translucent plastic. In 1946, Georama showcased an early illuminated plastic globe on the *Britain can make it* exhibition in the V&A Museum in London. In Italy, globe maker Rico pioneered illuminated plastic globes in the 1950s, soon followed by many others. By the 1970s, the globe industry had massively switched to plastic globes, with illumination as a standard option. At the end of the century, in millions of children's bedrooms around the world, the dark of the night was dispelled by the comforting glow of a plastic globe [nos. 20, 55, 158, 189].

Clarté
Metal sculpture carrying an illuminated glass sphere /
*Design Max Le Verrier /
France, 1928*

6

Bolis, the Italian publishing house from Bergamo, created this globe as a luxury object for interior decoration in the 1950s. The elegant twenty-four-centimeter illuminated glass sphere, with paper map gores, rests atop a sculptural stand of gold-colored bent metal tubes. The composition maintains a delicate balance, with a small metal ball at the end, reminiscent of the Moon orbiting around the Earth. The stand barely touches the sphere, creating a clever effect that enhances the impression of Earth floating in space. The gentle illumination casts a soft ambient light in the living room of the proud owner.

"Our creation the Electric Illuminated Globe, the modern lamp par excellence, is the great novelty of the day. It conquered the most elegant salons, where it puts a discreet note of bluish light on the bronze art pieces, as well as in the office of the businessman, happy to pursue his plans to conquer a world where light shines clear!"

Maryse Bastié (1898-1952), a French aviatrix who set numerous records for flight duration and distance, alongside a large globe adorned with a miniature airplane / *France, 1930s.*

E. Girard, 1933
Owner of Girard & Barrère
Paris globe maker

BAKELITE
GLOBES

A Material of a Thousand Uses

Belgian chemist Leo Baekeland (1863-1944) synthesized a new material in 1907 in the U.S., while searching for of an insulator for electricity. He baptized it Bakelite. It was a historic breakthrough: from that moment on, synthesized materials would go on to change the world. Their basic structure consists of long chains of polymers, which are malleable in any shape or form, hence the name 'plastics'. Thanks to its insulating properties, Bakelite was indeed useful to make electric sockets and light switches, but soon the material was used for all kinds of household objects and advertised as *"The Material of a Thousand Uses"*. The American Catalin Company acquired Baekeland's patent in 1928 and developed a comparable material called Catalin, available in many bright colors, an attractive material for combs, radios, buttons, or globes.

Thermoset

Not long after, globe makers discovered Bakelite. It was lightweight and malleable and could be mass-produced more efficiently than metal or wooden stands. It provided freedom of design, including rounded forms, which were fashionable at the time. Meanwhile, the spheres remained traditional, typically made of pressed cardboard, due to Bakelite's characteristic as a thermoset, a hard and smooth plastic prone to breaking upon impact. Rand McNally, the Chicago-based globe maker, was an early adaptor. In the 1940s it created the Pioneer, which featured a stand and horizon ring cast as one single volume of Bakelite [no. 168]. To enhance the appeal of the brown material to the public, the company marketed it as an expensive wood: *"The warm tones of the globe bleed with the rich walnut-finished bakelite [sic] stand and harmonize with any surroundings in which the globe is placed."*

Injection Molds

Leominster, Massachusetts was a place where many companies pioneered plastic manufacturing. In 1935, it earned the nickname Plastic Town. Commonwealth Plastics, a producer of plastic toys and houseware, acquired the first plastic injection molding machine, patented by the 'father of plastics', William M. Lester (1908-2005). The machine injects melted plastic into a cavity between two metal slabs. It reduces production time to a few seconds, essential to the success of mass-made plastic objects. In the 1940s, the company created the first full plastic globe, advertised as: *"The Wonder of Plastics - The only Globe of its kind in rich, durable, colorful plastic."* The technology facilitated the production of a relief globe of four and a half inches, cast in blue plastic, with a hand-colored political map on top. The art deco stand and meridian are also cast in plastic [no. 18].

Information Number Two
Catalog of Bakelite products / *General Bakelite Company / New York, 1912*

7
French-American designer Raymond Loewy (1893-1986), created an art deco globe-radio, entirely out of Bakelite in 1933. The cartography is painted in golden lines onto the brown Bakelite sphere, which conceals the radio receiver. Tuning and volume knobs are integrated into the bronze equator ring. The sphere is fixed by a single pin onto a hexagonal Bakelite base. It holds a loudspeaker inside and has slits to let the sound through. The object was promoted as a radical design: *"Here is radio in its proper setting: waves over the surface of the Earth - Here is design which fits into every home or office decoration."*

TRANSPARENT

Bubble Canopies

Chemist Otto Rohm (1876-1939) founded Rohm & Haas in Germany in 1907. The chemical company had a branch in Philadelphia, run by his partner Otto Haas (1872-1960), who had immigrated to the U.S. In 1933, Rohm & Haas synthesized a transparent acrylate, introduced to the market as a substitute for glass, called Plexiglas. As with many inventions, its initial application was military: a transparent bubble canopy for fighter aircraft. The seamless, impact-resistant, transparent dome improved safety and visibility for pilots. After the war, Plexiglas found numerous civil applications, such as transparent building materials and household items. When Robert H. Farquhar (1902-1983), an engineer and amateur astronomer, stumbled upon a Plexiglas bubble at a shipyard during the war, he was inspired to pursue his life-changing dream of creating a transparent globe.

Cockpit Bubble
Polishing Plexiglas
nose of attack
bomber at
Rohm & Haas
factory / *Bristol,*
Pennsylvania,
1940s

Plexiglas Spheres

Right after the War, Farquhar quit his job, sold his house to raise funds, and moved in with his mother-in-law. He trained six months in a Plexiglas factory and started a workshop in a Philadelphia garage. The Farquhar Transparent Globe Company quickly began manufacturing transparent Plexiglas globes, inspired by the concept of the airplane bubble. The fascinating globes resemble props from a science-fiction movie. Coastlines and borders, longitudes and latitudes are serigraphed in vivid colors on the sphere. As explained in the brochure, the globe was a great teaching tool: *"Marking crayons are provided so that any type of information may be put on the globe, such as, air currents, ocean currents, great circle routes, explorer routes, etc. The crayon can be easily wiped off with a soft cloth."* By looking through the globe, people on opposite sides of the world would better understand each other [no. 8, p. 66, p. 220].

Serigraphed Cartography

Rohm & Haas Reporter, the company's magazine, published an article in 1950 about Farquhar's Transparent Globe Company. It praises the manufacturing of transparent globes as a perfect example of the versatility of Plexiglas. The production process is described in a lot of detail. It involves a lot of precision: each hemisphere starts out as a circular flat Plexiglas sheet, onto which the cartography is serigraphed. The sheet is heated to three hundred fifty degrees Fahrenheit to soften the Plexiglas and form it into a half dome, using a vacuum press at forty-five pounds of pressure. Two hemispheres are then joined together with a seam to create a globe. The technology of a Plexiglas globe was a mayor twentieth-century breakthrough, enabling the production of unbreakable transparent globes in large quantities, which would become popular education tools.

8
Farquhar Transparent Globe Company in Philadelphia produced Plexiglas globes, terrestrial as well as celestial, ranging from four inches up to four feet in diameter, weighting twenty-two pounds. The base was sold separately and included options such as a ring, a Plexiglas cradle base, or a metal roller-tipped tripod base. Prices around 1960 varied from $29 for a fourteen-inch globe to $875 for a forty-eight-inch model. This twelve-inch tectonic model from the 1960s, shows mountains, rivers, and fault lines in the oceans. The transparency allows observation of how the fault lines continue all around the Earth.

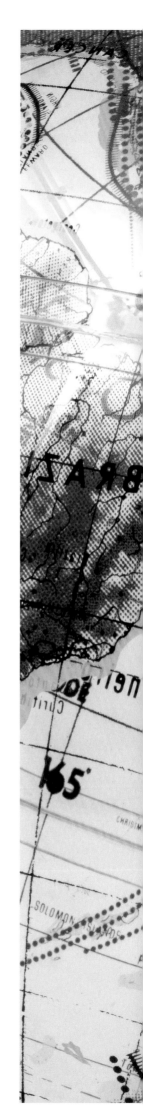

PLASTIC
GLOBES

Elephants and Billiard Balls

American inventor John Wesley Hyatt (1837-1920) sought a substitute for ivory. One day in 1869, he concocted celluloid. It was used for false teeth, piano keys and film rolls, and promoted as nature's savior: *"... no more elephants have to be killed to make billiard balls!.."* Throughout the twentieth century, synthetic materials replaced metal in cars, wood in furniture, glass in packaging, cotton in clothing, and silk in stockings. Plastic was welcomed as a utopian material, safe and hygienic, sturdy and flexible, moldable in any shape. In 1942, an observer wrote: *"... the word plastics is glamorized as a mystery and the materials themselves are declared to be the evidence of a chemical revolution so far-reaching that our very lives might be altered."* By the end of the century, the dream had turned into a nightmare: plastic waste emerged even in the most remote corners of the planet.

Plastic Revolution

Plastic revolutionized the globe industry. A traditional globe required a lot of different materials, such as a papier-mâché, plaster, printed paper, varnish, metal, and wood. From the 1920s onward, celluloid was occasionally used as an alternative material for spheres. In the 1930s, Bakelite served as a substitute for wooden or metal stands. In the 1950s, acrylate spheres were molded and covered with paper map gores. In the 1960s, the full plastic globe was born: cartography was printed directly on a flat acrylic sheet, heated and pressed into a vacuum mold to form hemispheres, then glued together at the equator and mounted on a plastic base. Since the 1970s, plastic globes are prevailing as the industry standard, representing the final stage in the transition from expensive handmade globes to cheap mass-produced objects, while paper globes have once again become niche products.

Plastic versus Paper

Globe makers who adhered to traditional paper globes risked going out of business. In the second half of the twentieth century, companies like Felkl in Prague, Taride in Paris, or Weber Costello in Chicago closed down. They lost the competition to dynamic globe makers who had made early investments in plastic molding machines and began producing plastic globes, such as Replogle in Chicago or Columbus Verlag in Stuttgart. New companies, founded after the Second World War, such as Rico in Florence, had a head start as they immediately specialized in plastic globes, not hindered by a tradition of paper globes [nos. 9, 19, 20, 124, 140, 145]. Today, only a handful of European and American globe-making companies remain. Together with a few Asian manufacturers, they constitute the globe industry of the twenty-first century, producing millions of generic plastic globes each year [no. 189].

The Most Amazing Wrap Ever Developed!
Ad for Saran-Wrap plastic / *Dow Company / U.S., 1953*

9
Rico in Florence, a young Italian company, began manufacturing plastic globes in the 1950s. It gave them an advantage over the traditional globe makers who had to transition from traditional production methods. Rico's globes consist of two plastic hemispheres formed in a mold.
The map is printed on a thin plastic foil, and then pressed onto the hemispheres. The meridian and tripod stand are also made of plastic. The globes came in various colors, shapes and sizes, such as this twenty-five-centimeter model. Cheap illuminated plastic globes proved serious competition for the expensive glass globes.

INFLATABLE

Blow-Up Furniture

American chemist Waldo Lonsbury Semon (1898-1999) synthesized vinyl for the B.F. Goodrich Company in 1926. Polyvinyl chloride or PVC is the second most-produced plastic in the world today. Its first applications were golf balls and shoe heels. Elastic and airtight, it is perfect to create inflatable products, such as the brightly colored beach ball, which appeared on Californian beaches in the late 1930s and soon became an icon of surf culture. During the 1960s and 1970s, when a free-floating lifestyle was popularized, brightly colored inflatable vinyl furniture conquered the living room such as the Aerospace Collection, designed by French-Vietnamese engineer Quasar Khanh (1934-2016), nicknamed 'The Architect of Air'. Pop art architects such as Haus Rucker Co. in Vienna, Archigram in London or Archizoom in Milan designed inflatable buildings.

Blow-Up
Jane Birkin sitting in a Quasar Khanh inflatable vinyl chair /
Advertisement, 1969

Vinyl Globes

The first inflatable vinyl globes hit the market in the early 1950s. Contrary to what one would expect, they were not marketed as cheerful toys but as serious geographical instruments. The printing technique on vinyl allowed for sharp and detailed cartography, making inflatable globes strong competitors to traditional globes with paper maps. The inflatable globes came in various sizes and with various metal globe stands, including both floor and table models. Once deflated, a lightweight vinyl globe could easily be folded and carried around, a big advantage over the transportation challenges posed by the heavy and bulky metal and wooden globes. No wonder that inflatable vinyl globes were promoted with the same slogan as The Blow, the famous inflatable seat of the 1960s: *"... easy to blow up, easy to transport, and easy to store ...."*

Polar Valves

The construction of inflatable globes is similar to that of a beach ball. Twelve vinyl map gores are fused into a sphere. Valves at the poles serve to inflate the sphere and make it spin in its lightweight metal stand. Various globe makers released inflatable globes, such as Georama in London, Taride in Paris (in collaboration with rubber boat producer Sevylor), Ravenna in Berlin (in collaboration with Japanese Toshiba) or Michael Seidel in Zirndorf. Hammond in New Jersey even produced an inflatable globe with a lamp inside, which seems rather dangerous given vinyl's flammability [no. 161]. Inflatable globes were great for advertising, Pan Am or Air France had their logo printed on it to evoke dreams of exotic beaches and encourage the purchase of tickets [nos. 15, 16, 17]. However, by the end of the century, inflatable globes had lost their serious appeal and were once again relegated to being used as beach balls.

10

Standard College Globes Inc. in New York produced this inflatable globe in the 1950s, called Duke. The ten-inch vinyl sphere has a detailed multi-colored cartography, aimed at educational use. The sphere has valves at its poles for inflation and attachment to the stand. The valves are used as hinges, so the vinyl ball can spin around its axis, just like the Earth. The meridian and base are constructed from bent steel wire, providing a functional lightweight structure that can be disassembled for easy transport, together with the deflated sphere.

RELIEF

Heights and Depths

In 1953, Australian explorer Edmund Hillary (1919-2008), and Nepalese mountaineer Tenzing Norgay (1914-1986) became the first people to reach the summit of Mount Everest, the highest point on Earth. Seven years later, Belgian explorer Jacques Picard (1922-2008) descended in his bathyscaph to the Mariana Trench, the deepest point on Earth, more than ten thousand kilometers below sea level. In the twentieth century, heroic ambition, scientific curiosity, and technical progress brought man to the most extreme places on Earth. For cartographers, it was a challenge to visualize these heights and depths on a map. At the beginning of the twentieth century, standardization of graphics and colors was not yet fully developed. The manufacturing of realistic three-dimensional relief globes was an even more painstaking manual process, eagerly awaiting innovation.

Schattenplastik und Farbenplastik

Austrian cartographer Karl Peucker (1859-1940), in his book *Schattenplastik und Farbenplastik* of 1898, describes how to evoke the illusion of three-dimensional heights on a map. The first method, Schattenplastik, uses hand-drawn shadow hatching of mountain ranges. In the 1960s, globe maker Hammond photographed the actual shadows of geographical scale models and incorporated them into map drawings. The second method, Farbenplastik, utilizes a color system to represent heights: water is blue, plains are green, hills are yellow, and mountains are orange to red. Since the human eye perceives reddish colors as more foregrounded, the colors create an illusion of height. It became the standard method. Over the course of a hundred years, we have become so accustomed to this color system that we now instinctively interpret the heights without having to consult the legend.

Raised Relief

Raised relief globes were originally developed for the blind. Papier-mâché or plaster spheres were cast in molds with exaggerated relief and manually painted. The same technique was used later to pour plastic relief globes [nos. 22, 23]. In the 1920s, Leipzig globe maker Paul Räth simplified the process by stretching wet paper map gores over the relief spheres, resulting in a somewhat distorted appearance [no. 51]. In the 1950s, Rico in Florence patented a machine to produce plastic relief globes. The cartography is printed on flat sheets of plastic, which are then heated and vacuum pressed into a negative relief mold. Two or four curved sections are combined to form a relief sphere, a process that would become the standard [nos. 21, 145]. Replogle emphasized the educational value of its relief globes: *"You can feel the mountains! Understand better how elevations create climates, affect populations, boundaries, and transportation."*

Every inch a real smoke! Climber Barry Corbet on the Mount Everest ascent / *Camel Tobacco ad, 1963*

11

Weber Costello, based in Chicago, developed a simple method to create relief globes in the late 1940s. The True Vue Globe is an existing model featuring a ten-inch sphere, with paper map gores displaying a political map with black oceans. Mounted on top of the paper sphere are two relief hemispheres cast in blank Lucite. This transparent plastic outer layer allows the paper map underneath to remain visible. The two layers blend together, creating the illusion of a relief globe without the complexity of producing a colored relief map. The globe has a metal meridian and a wooden cross base.

More Metal Globes

12

This small tin globe, made by Chad Valley, a Birmingham toy factory, also exists under the brand logo Reliable Series. The 1950s model was available in different variations. In this case, the vividly colored cartography on the ten-centimeters lithographed tin sphere makes a sharp contrast to the black-lacquered metal stand and folded metal meridian.

13

Replogle in Chicago made a line of globes for children, parallel to their more sophisticated globes. This model of the 1950s has a six-inch lithographed sphere with a detailed cartography, displaying the U.S. states in different colors. It is fixed to a graded meridian and a half-sphere base, both in stark red, an energizing contrast to the electric-blue oceans.

14

Ohio Art Company in Bryan, Ohio, made this children's globe in the 1930s. The six-inch metal sphere is fixed to a graded full meridian. The octagonal metal base is lithographed with a fake wood veneer print, zodiac signs and a calendar. The political map shows recent events, such as the flight routes of Lindbergh and the Graf Zeppelin.

More Inflatable Globes

15

URB Plastics in New York, produced swim rings, but also this eight-inch globe for Pan American Airlines in the 1940s. It is the first-known inflatable globe made of vinyl, in this case Koroseal, developed by F.B. Goodrich. The cutting pattern results in a flattened sphere. The pictorial cartography shows air distances, animals, flowers, and hurricanes.

16

"Fly Pan Am Jets Around the World" is the title on the cartouche of this 1950s inflatable advertisement globe, copyright of the General Graphics Corporation. The twelve-inch model has a detailed physical map, but no indication of air routes or exotic touristic features. It spins on its air valves in a demountable steel wire tripod.

17

Parisian globe maker Taride created this globe in the 1960s, in collaboration with rubber boat producer Sevylor. The model was used for advertising purposes by Air France. Twelve printed vinyl map gores are fused together to form a twenty-eight-centimeter sphere. The air valves at the poles enable the globe to spin on its steel wire stand.

More Plastic Globes

18

Commonwealth Plastic Corp. in Leominster, U.S. was a pioneer in plastic injection technology. In the 1940s, it produced the first full-plastic globe using a type of Bakelite. The four-and-a-half inch sphere has a raised relief, molded in fluorescent-blue plastic. The colored map is hand painted. The stand and meridian are also crafted of molded plastic.

19

Danish Scan-Globe, a subsidiary of the American globe maker Replogle, developed various plastic globes in the 1970s. This twenty-five-centimeter model features shiny plastic, vivid colors, and a folded Plexiglas stand. The contemporary design transforms the scientific instrument into a decorative object, a true representative of its time.

20

Danish Scan-Globe patented this Spot-Globe in 1975. By adjusting two graded knobs to the latitude and longitude of any place on Earth, an internal spotlight marks its location. The twenty-seven-centimeter illuminated sphere is made of two plastic hemispheres with the map printed on them. The circular base is also made of plastic.

More Relief Globes

21

Nystrom in Chicago, U.S., manufactured relief globes for school education in the 1990s. The globe utilizes a double system to depict the natural heights of the Earth: its physical map is colored following the Farbenplastik method. The fourteen-inch plastic sphere features an embossed relief. It is mounted on a gyroscopic meridian and metal base.

22

German globe maker Jörn Seebert in Staufen im Breisgau created large relief school globes in the 1950s. The fifty-centimeter relief sphere is made of polyester resin. The relief is exaggerated by fifty to one to augment the effect of heights. The sphere rests on a single pin of the metal base, allowing blind students to examine its surface.

23

German road map maker JRO in München produced this Braille globe in the 1960s. Mountain ranges, latitudes, and longitudes are cast in relief using resin. Cities and rivers are also given height, allowing users to feel their positions. To most important geographical names are indicated with dots in Braille script on the tactile thirty-centimeter globe.

VOGUE

BEAUTY...on a 1943 working schedule

ADVAN
RETA
TRAD
EDITI

Fashions and
facts for
YOUNG
MOTHERS

STYLE

World Peace

Local styles transformed into global fashion during the twentieth century. A driving force behind this transformation was the widely distributed international magazines that spread the latest fashion in full color offset print, such as the glossy *Vogue*. In November 1943, in the heat of the Second World War, *Vogue*'s mood was still rosy. The theme of the month was fashion and facts for young mothers.

On the cover is a picture of model Susann Shaw (1929-1978), gazing through a transparent globe. The photograph was taken at the *Airways to Peace* exhibition, which was running at the time in the Museum of Modern Art in New York. The introductory text of the exhibition explained how air travel had globalized the world: *"The modern airplane creates a new geographical dimension. A navigable ocean of air blankets the whole surface of the globe. There are no distant places any longer: the world is small and the world is one. The American people must grasp these new realities if they are to play their essential part in winning the war and building a world of peace and freedom. This exhibition tells the story of airways to peace."*

Among many maps and globes, it featured this Antipode Globe, a transparent, glossy glass sphere with continents painted in vivid colors. It illustrates the relationship between places on opposite sides of the globalized world, showing how they are connected within a few hours of flight.

A globe is not only a scientific instrument but also a decorative object that reflects the cultural context and the spirit of its time. In the twentieth century, the design of globes adapted to the prevailing architectural styles, from art nouveau to art deco, from modernism to postmodernism. By the end of the century, all styles had merged together, resulting in the generic globe.

ART NOUVEAU

Plants and Lashes

British designer William Morris (1834-1896) led the arts and crafts movement in the late nineteenth century. He decorated textiles, furniture and household objects with intricate vegetal patterns. Morris was also a utopian socialist who aspired to create a classless society and elevate the status of craftsmanship. This position was paradoxical at a time when handcrafted objects became costly while industrial products became less expensive. Around the turn of the century, the art nouveau style emerged, building on Morris' aesthetic principles, but not his political ideas: art nouveau mainly catered to a bourgeois urban elite. Architects utilized lightweight steel constructions and large glass panes, decorated with plant motifs or so-called 'whip strokes', that wind their way up in asymmetrical compositions. This set the stage for the modernism of the twentieth century.

Elegant Curves

Belgian architect Victor Horta (1861-1947) was a pioneer of art nouveau, designing exuberant mansions and public buildings in a style that rapidly spread over Europe. In Paris, the streets were adorned with the iconic art nouveau entrances of the Paris Métro. Known as the Sezession in Vienna and Jugendstil in Germany, named after the progressive art magazine *Die Jugend*, the style flourished in various forms throughout the continent. Art nouveau incorporated not only architecture but also various objects, including globes. The bulky wooden globe stands and cast-iron tripods, heavily decorated in historical styles, were replaced by elegant art nouveau stands that were slender and curvy, made of metal, and adorned with simple vegetal lines. Interestingly, the most striking examples were not created by European globe makers, where art nouveau originated, but by their American counterparts.

British Map Gores

J.L. Hammett Company, a school supplier from Boston, created an art nouveau floor globe in the 1910s. Its stand featured three curved legs, reminiscent of the furniture designs of that era. The map gores were made by G.W. Bacon (1830-1922), a map maker from London. Since drawing maps was a specialized and expensive task, it was common practice for early twentieth-century American globe makers to purchase printed map gores from British map makers. The young New Jersey globe maker C.S. Hammond & Co. bought its map gores from Scottish firm A.W. & K. Johnson, and simply glued its own brand name over the original. In the 1920s, Hammond created a number of elegant globes that rested on art nouveau-styled cast-metal stands, characterized by slender curvatures and vegetal lines. Often, a few strong details distinguish these globes, such as a meridian ending in a leaf form.

Art Nouveau Interior
Stairs of Hôtel Tassel / Architect Victor Horta / *Brussels, 1894*

24
C.S. Hammond & Co., based in New Jersey, created this striking example of an art nouveau-styled globe in the 1920s. The slender cast-metal frame is designed in a single asymmetrical movement from base to top, holding a delicate six-inch plaster sphere that balances off the South Pole, tilted to suggest Earth's natural inclination—a powerful suggestion of Earth floating in space. After one hundred years, the metal stand has developed a greenish copper-oxide patina that blends well with the faded greenish tones of the paper map gores, and their shiny brownish varnish.

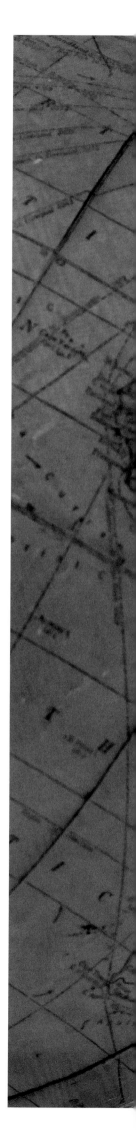

ART DECO

Black Square on a White Field

In 1915, Russian painter Kasimir Malevich (1879-1935) hung a painting of a black square on a white background in a corner of the exhibition room in Petrograd. This work marked the beginning of Russian suprematism, an abstract art movement characterized by rudimentary forms, colors, and compositions. Abstract art emerged in various parts of Europe during the 1910s. In Paris, the rise of cubism focused on squarish forms, while in the Netherlands, the De Stijl group created radical rectangular compositions using primary colors. Italy's young artists, fascinated by machines and speed developed futurism. Meanwhile, German artists practiced expressionism, characterized by dark lines. African art, with its strong colors and patterns inspired many Western artists. After the First World War, elements of these avant-garde movements merged into a popular mainstream style called art deco.

Decorative Compositions

The breakthrough for art deco came at the 1925 *Exposition des Arts Décoratifs* in Paris, which is also where it received its name, in retrospect. The style was characterized by a hodgepodge of squares, circles, and triangles, along with bright colors and expressive patterns. During the interwar period, art deco became the leading bourgeois style, influencing furniture, clothing, buildings, and even globes. European globe makers such as Paluzie in Spain or Girard Barrère in France transformed their traditional wooden globe stands into lightweight bases of folded metal [nos. 34, 35, 36]. The sphere is often only fixed to its base by a pin, aligning with the Earth's natural inclination. It turns the sphere and base into a geometric sculpture of triangles, circles, or squares. A notable example is a small art deco globe made by British firm George Philip & Son, featuring a sphere on a gold-painted, triangular cast-iron base [no. 37].

Modern Typography

Art deco globes featured fresh and modern-looking stands, while still retaining traditional spheres made of pressed cardboard and paper map gores. However, the cartography was redesigned to improve readability, using fresh pastel color schemes and modern typography. Graphic features, such as curly ocean currents, enhanced the overall composition [nos. 3, 25]. Globe makers of the period also updated the geographic features of their globes, reflecting the latest political and economic changes, such as the new European borders after the First World War and new countries like Manchoukuo, a part of China occupied by Japan. American globe makers started to introduce media events on their maps, highlighting significant achievements like Charles Lindbergh's (1902-1974) transatlantic flight, the polar explorations of Richard E. Byrd (1888-1957), and the first intercontinental services of the Zeppelin airships.

Art Deco Interior
Greta Garbo on the set of *A Woman of Affair* / Designer Cedric Gibbons / Hollywood, 1928

25
Perrina, a globe maker in Paris led by director Picquart, created elegant globes in the mid-twentieth century. One fine example of their popular art deco-styled creations, produced during the 1950s, features a twenty-centimeter sphere covered with beautifully colored paper map gores, adorned with dynamic swirls of ocean currents in refined cartography. The cartouche includes a list of abbreviations for the names of the colonial powers, which at the time still occupied large parts of the world. The metal base is a stepped-shaped cone, a formal composition often used in art deco design.

STREAMLINE

Illusion of Speed

French designer Raymond Loewy (1893-1986) immigrated to New York in 1919, during the city's skyscraper boom. Ever taller towers rose up, in art deco styles, such as the Chrysler Building or the Empire State Building. Loewy opened his own design studio and became an important precursor of American art deco which he single-handedly transformed into the streamline style. The style is characterized by dynamic lines and curved shapes that reinforce the suggestion of speed. Above all, it expresses a futuristic, machine-age American lifestyle. This popular style, prevalent before and after the war, was applied by Loewy to trains and cars, kitchens and radios, and even the interiors of American spaceships. Not a single object escaped his streamlined makeover, even those not meant to fly, such as pencil sharpeners or globes [no. 7].

Floating through Space

American globe makers picked up the streamline style. They continued to produce cardboard spheres with paper maps but replaced their wooden stands with streamlined metal mounts. By that time, globes had transformed from scientific instruments to decorative objects for middle-class homes. Consequently, it became commercially important to create globes that matched the aesthetics of interiors and featured interesting elements, serving as conversation pieces for guests. Just as cars came with streamlined hoods, globes appeared with cast-aluminum bases, shaped like abstract wings, water surfers, or even stylized airplanes. The inclined stands created the illusion of a planet moving through space at high speed [no. 26]. The streamlined grey metal bases were often combined with cartography featuring black-colored oceans, enhancing the machine aesthetics.

Airplanes and Wings

In the 1930s, Replogle from Chicago created a number of streamlined globe stands. One model, appropiately called Streamlined features an elaborate cast-aluminum base that gives the impression of the sphere cutting through water like a speed boat [no. 38]. A metal half-meridian rises up from the composition. Speed was also the focus of the globe's advertisement: *"Follow the world's swiftly moving events with this 10-inch globe!"*. Another Chicago globe maker, Weber Costello Company, also created streamlined globes at the time. Some of its models have a shiny metal stand in the shape of an airplane, leaning towards a kitschy aesthetic [no. 39]. A more subtle model is a black ocean globe that seems to hover above an abstract rounded-off V-shape. The cast-metal base could be interpreted as a stylized bird, flying through the sky, creating another illusion of speed.

Streamlined Train
Pennsylvania
Railroad S1
Locomotive /
Designer Raymond
Loewy / U.S., 1939

26
Around 1935, Replogle in Chicago advertised its New Starlight illuminated globe with bold claims: *"MODERN, DARINGLY DIFFERENT"* and *"MODERN AS TOMORROW!"* The ten-inch illuminated glass sphere with dark paper map gores rests on a cast-aluminum base. Its smooth rounded-off shape seems to refer to the ancient symbol of the 'winged wheel', the attribute of Hermes, the fast messenger from Greek mythology. This design suggests the globe is flying through the universe at great speed. The sphere is fixed to the central axis at the South Pole and is embraced by an asymmetrically meridian on both sides.

COLORED OCEAN

Black Water and Blue Ice

Swiss artist Johannes Itten (1888-1967), was a teacher at the Bau-
haus, the famous German school for modernist design, in the early 1920s. He developed a color theory, based on the relationship
between primary, secondary and tertiary colors, which was widely
embraced. Color is one of the most challenging design elements,
especially when used to convey information, as is the case in cartography or packaging. Colors lack a universal meaning, which be-
comes clear when one visits a supermarket abroad: a milk package might be blue in one place, but red in another. The same goes
for map colors. Globes often use blue to represent water, but this
is not a general convention. Geographical names even suggest
color variations, like Red Sea or Black Sea. Frozen water might
appear white, but ice is rarely shown on globes to demonstrate
that the Arctic is a sea and the Antarctic is land.

A Globe in Every House!

Globe makers in the 1930s tried to open up a market for middle-class families. Replogle in Chicago and Columbus Verlag in
Berlin used the same slogan: *"A globe in every house!"* To make
globes fit into the living room, map colors followed popular taste
rather than geographical convention. Since oceans cover two-
thirds of the Earth, ocean coloring defines the overall impression.
Traditionally, blue was used to color the oceans, but some globe
makers painted the oceans silver, black, red, or yellow to create a
stunning look [nos. 26, 27, 72, 170, 176]. This trend created a hype,
as Cram's director Edward A. Peterson (1882-1966) recalled in an
interview: *"No sooner had these colored ocean globes been placed on
counters, quite a number of women, casting about for a conversation
piece and a mild touch of culture, uttered small shrieks and grabbed
globes like mad. Anyway, it got the globe into the living room."*

Black Oceans

Most American globe makers praised the beauty of their black
ocean globes. Rand McNally promoted its Midnite model with the
slogan: *"Dramatic ... highly as in a rich dark-marine-blue framing
the deep pastel shades of the land areas"*. The black ocean trend
persisted for three decades, but it never crossed the Atlantic. In
Europe, globe makers kept their oceans blue. George F. Cram
even created a Cherry Red Ocean Globe, with red oceans, as if
Earth were an exotic hot planet. The transformation of the globe
from a scientific instrument to a decorative object was promoted
in Cram's 1953 dealer catalog: *"A NEW IDEA THAT HAS CAUGHT
ON - Decorative furniture-style mountings and colorings. Don't over-
look these beautiful models. They step up the eye appeal of your display and are steady sellers to discriminating buyers"*.

**Color Sphere in 7
Light Values and
12 Tones** / Designer
Johannes Itten /
Germany, 1920s

27

George F. Cram Company from Indianapolis
launched a line of Silver Ocean Globes, in
the mid-1930s. The oceans were printed in
a silver-metallic color, and there was even a
Golden Ocean Globe, which had a golden-
metallic shine. Cram's experiments with
silver and gold only lasted for a few years,
as the metallic ink was technically difficult
to print. This twelve-inch model featured
a full metal meridian and a rounded base.
The effect of a silver sphere set in a black
ring gave the globe a sophisticated look,
that certainly matched well with the
contemporary art deco interiors of the
period.

SCULPTURE

Greek Gods

German-American sculptor Lee Lawrie (1877-1963) created the iconic bronze sculpture in front of the Rockefeller Center in New York. The monumental forty-five-foot *Atlas* was inaugurated in 1937. According to Greek mythology, Atlas was punished by Zeus and sentenced to carry the universe on his shoulders. This theme has been used in many artistic representations since ancient times. The oldest globe still in existence is part of a Roman sculpture called the *Atlas Farnese*, dating from around 150 AD. In this sculpture, Atlas carries a sixty-five-centimeter marble celestial globe, depicting forty-one chiseled constellations. In American art deco architecture as well as in the neo-classical architecture of Nazi Germany and the Soviet Union, heroic sculptures were often integrated, a fashion that spilled over to globe stands of that period.

Conversation Pieces

During the art deco era, popular models featured an illuminated glass globe supported by an elegant bronze sculpture. These luxury globes were intended as conversation pieces in affluent American households, designed to complement art deco interiors. In Cram's 1938 catalog, this concept was openly promoted as a selling feature for its Dolphin Globe: "... *Visitors will comment favorably on its distinctive appearance....*" In this particular model, an illuminated glass sphere is supported by a bronze sea monster, which makes an obvious nod to the sea monsters found on many classic globes from times when sailors encountered unknown animals in remote places. The globe sculpture is not very dolphin-like: it more closely resembles a sea snake with fish scales and an immense tail that wraps around the sphere to morph into a meridian.

Atlas Base

Sixteenth-century Flemish cartographers Mercator and Ortelius named their book of maps *Atlas*, in honor of the Greek god. Since then, it remained the name for any book of maps. The limited surface area of a globe restricts its geographical information, whereas the maps in an atlas provide ample space for detailed cartography and various scales. This prompted American globe makers in the 1930s to produce 'atlas base' globes, featuring a wooden book casing as a base to house an additional atlas. Combining a globe with an atlas is both scientifically ingenious and commercially advantageous, as it allows publishers to sell two products simultaneously. To promote this concept, Replogle advertised its Atlas Base Globe with swagger: "*The most fastidious person would be pleased to own this globe ... always ready for instant reference.*"

Atlas
Bronze sculpture in front of the Rockefeller Center / Sculptor Lee Lawrie / *New York, 1937*

28
George F. Cram Company from Indianapolis created this striking globe around 1940. It features a ten-and-a-half-inch illuminated glass sphere, supported by a bronze Atlas sculpture. In an advertisement, Cram confirms the sculpture is a copy: "*What could be more appropriate for a globe mounting than Atlas supporting the world on his shoulders? The inspiration for this distinctive work of art is the famous Lee Lawrie statue of Atlas which stands in the forecourt of the International Building at Rockefeller Center, New York City.*" Unfortunately, this is a mistake: the Greek god Atlas did not carry Earth, he carried the universe.

"No sooner had these colored ocean globes been placed on counters, than quite a number of women, casting about for a conversation piece and a mild touch of culture, uttered small shrieks and grabbed globes like mad. Anyway, it got the globe into the living room."

Edward A. Peterson, 1941
Owner of George F. Cram Company
Indianapolis globe maker

Photo shoot
for new spring
fashion / Elegant
dressed lady
admiring globes
in a showroom /
U.S., 1953

MODERNIST PLYWOOD GLOBES

Molded under Pressure

Finnish architect Alvar Aalto (1898-1976) experimented with plywood furniture in the early 1930s. The plywood technique, developed in the late nineteenth century, allows for the production of inexpensive wooden objects, while offering structural strength and freedom of form. The versatile lightweight material was initially used to build small boats and airplanes. It consists of thin layers of wood veneer laminated together to form strong wooden panels. Molded under high pressure, plywood can be formed into any desired form. Aalto used curved plywood for his iconic modernist Paimio Chair, whose seat, back, armrests and legs are all shaped out of a single piece of bent plywood. In the decades that followed, various Nordic designers successfully developed plywood furniture, making it a popular element in mid-century modernist interiors.

Wooden Cross

American globe makers started experimenting with wood technology during the Second World War, when metal was prioritized for the military. In the mood of the period, it made sense to abandon the frivolous art deco metal stands for more sober wooden stands. Unlike the traditionally decorated turned wooden stands, they used two wooden planks, slid them together to form a cross, and placed a loose sphere on it, creating the so-called 'cradle base'. The edges were left exposed, following the functionalist approach of using plain material without decoration. The cradle was lined with felt, allowing the globe to be rolled without scratching and picked up to observe all sides. Wooden cradle globes, being cheap and robust, became popular American school globes. Several models were available, featuring simplified map images tailored to the various cognitive levels of school children.

Freedom, Victory and Liberty

Chicago school supplier Denoyer-Geppert launched a series of wooden globe bases in the 1940s, adorned with patriotic names that anticipated the hoped-for outcome of the war. The Freedom was a low cradle mount that offered maximum visibility to the sphere [no. 184]. The Victory rested on an asymmetrical wooden cross [no. 163]. The Liberty was a cradle mount with a wooden horizon ring [no. 178]. The robust orbs are covered with paper map gores developed by Levinus P. Denoyer (1875-1966), who created the so-called Cartocraft Globe. The straightforward and clearly readable cartography of the physical-political map with its strong contrasting colors blends nicely with the wooden base. The no-nonsense design reflects the mindset of the era, emphasizing a break from the past and a move towards a modernist future.

Paimio Chair
Plywood armchair
for the sanatorium
in Paimio /
Architect Alvar
Aalto / *Finland,*
1932

29
SVH, a Dutch company based in The Hague, originally manufactured cardboard boxes but expanded to office stationery in the early twentieth century. In the 1950s, the company distributed Dutch-language globes under its own brand name, produced by Columbus Verlag in Germany. This clever model features a thirty-three-centimeter sphere with paper map gores. It sits on an elegant stand made from bent plywood, in the spirit of the mid-century modernist furniture. The four-legged plywood base supports a graded plywood horizon ring and a metal meridian.

MODERNIST STEEL WIRE
GLOBES

Mid-Century Modernism

Los Angeles architect couple Charles (1907-1978) and Ray Eames (1912-1988) were famous designers in the mid-century modern period. Right after the war, they pioneered a revolutionary technology for furniture design: mounting lightweight plastics seats on thin steel wire frames. When the Museum of Modern Art in New York organized the Low-Cost Furniture Design Competition in 1948, Charles and Ray Eames submitted La Chaise, a groundbreaking design featuring a fiberglass tub floating on steel wire legs. The iconic concept would remain a bestseller throughout the twentieth century. The couple designed a number of other models following the same principle that are classics today, such as the DAR Chair. Various other designers created similar chairs. Harry Bertoia's (1915-1978) Diamond Chair even featured a seat made entirely of steel wire.

Less is More

American and European globe makers found inspiration in the modernist style and began creating globe models with spheres floating on steel wire stands. These straightforward models represented a clear break with the pretentious art deco globes. Rather than impressing with heavy sculptures, the new designs proposed a light-footed functionalism, following the 'less is more' principle. Steel wire was bent to form tripods, sledges, grills, or ring-shaped globe stands. The spheres were industrially produced of metal, cardboard, glass, or plastic. The cartography was revamped, using brighter colors and contemporary fonts that were clear and easy to read. These radical changes transformed globes into stark modernist objects that reflected the optimistic feeling of the postwar era, announcing a better world after the dark times of devastation.

Playful Details

Replogle from Chicago created steel wire models after the war. Its eight-inch lithographed tin Simplified Globe features a bare modernist design with tong-in-cheek details such as a wing nut to fix the metal meridian or a rack of braided wire, reminiscent of kitchenware. Its four feet have playful red plastic toes to prevent scratches on modernist wooden side tables [no. 40]. European globe makers like Dalmau in Spain or Paravia in Italy released various globe models supported by steel wire tripod stands. French company Perrina created an elegant, illuminated glass sphere on a steel wire sledge [no. 41]. In Paris, Girard, Barrère & Thomas made joyful globes with steel wire bent into tripods, racks, or rings, and even steel wire meridians [nos. 42, 138], while Taride made stands of three steel wire pins with rubber balls on the ends [nos. 137, 157].

DAR Eiffel Chair
Plastic seat on a
steel wire base /
Design Charles and
Ray Eames /
U.S., 1950

30
George F. Cram Company from Indianapolis had various globes in its catalog during the 1950s that rested on modernist steel wire stands. This celestial globe consists of a twelve-inch sphere with paper map gores, supported by V-shaped steel wire legs welded onto a steel wire ring. It features a metal horizon measure ring similar to those used in older Cram's models [NO. 165]. The slender stand is painted in a dull bronze, which nicely contrasts with the stark-blue star map. The cartography follows the modern international astronomical standard by depicting constellations as rectangular areas.

MODERNIST FLOOR
GLOBES

Architects and Centerfolds

Chicago copywriter Hugh Heffner (1926-2017) founded *Playboy Magazine* in 1953, the famous magazine for men that contributed to the transformation of postwar American culture. In this male-dominated period, it combined glossy pictures of naked women with short stories by famous writers and interviews with influential intellectuals. As a typical example, a 1961 article shows six men in suits and ties, in stark contrast to the nude female centerfold a few pages further on. The catchphrase reads: *"Unfretted by dogma, the creators of contemporary American furniture have a flair for combining functionalism with esthetic [sic] enjoyment."* The six men are the American design avant-garde: George Nelson (1908-1986) Edward Wormley (1907-1995), Eero Saarinen (1910-1961), Harry Bertoia (1915-1978), Charles Eames (1907-1978), and Jens Risom (1916-2016).

Sober Elegance

The traditional model for a gentleman's study room was a heavy floor globe, a piece of decorated furniture, handmade by a skilled cabinet maker. In contrast, most modern globes were lightweight table models that could be placed on a school desk or a living room sideboard. However, the idea of a free-standing status-boosting globe did not completely disappear. During the mid-twentieth century, several furniture designers created luxury modernist floor globes. Danish-American architect Jens Risom (1916-2016), co-founder of Knoll Furniture in the 1940s before launching Jens Risom Design, created a modernist floor globe for Replogle. The wooden stand, supporting a blue ocean sphere, exudes the sober elegance of Scandinavian design: four tapered legs on wheels connected to a central cross that supports an open horizon ring.

Design Consultants

The introduction of an independent design consultant was a novelty. It resulted from the division of labor in the process of mass production, where the work of the traditional craftsman was distributed among various specialists and machines. The 'industrial designer' was brought in to coordinate the design and optimize the production by integrating commercial, technical and aesthetic aspects. One example is Edward Wormley (1907-1995), who modernized the Dunbar furniture collection in the 1930s to fit contemporary taste. Although not as avant-garde, he was famous for blending classical design principles with modernism. Wormley created an elegant floor globe for Dunbar called the Cosmopolitan. The illuminated sphere is mounted on a brass meridian, fixed to a tall walnut tripod floor stand that supports a horizon ring with three tapered arms.

Unfretted by Dogma
Avant-garde architects sitting on their own designs / *Playboy Magazine, July 1961*

31

Adrian Pearsall (1925-2011) was an American furniture designer whose company Craft Associates made modernist furniture in the 1950s and 1960s. He collaborated with Replogle to create this modernist floor globe, featuring four curved legs in dark walnut that embrace a brightly colored illuminated sphere. The sixteen-inch sphere is made of plastic, covered with paper map gores, and set in a brass full meridian. The globe evokes the feeling of the atomic age. The elegant composition, fit for a modernist interior, is reminiscent of the sculptures of Henry Moore (1898-1986).

MODERNIST LUXURY
GLOBES

Leather and Walnut

The Los Angeles architect couple Charles (1907-1978) & Ray Eames (1912-1988) not only developed low-cost mass-produced chairs but also designed luxurious modernist-styled furniture. These high-end models are based on similar design principles, such as the efficient prefabrication of elements and the use of bent plywood or steel legs. The main differences are the precious materials and luxurious finishings, such as mahogany veneer, thick leather upholstery, or chromed steel. The chairs were designed for director's offices or gentleman's studies. Similarly, there was a niche market for luxury globes, even as mass-produced models became popular. A luxurious modernist globe was considered an object of prestige on a director's walnut desk or a gentleman's mahogany side table. It characterized the owner as a powerful man of the world, a connoisseur of good taste.

State and Business Leaders

West German globe maker Columbus Verlag, which manufactured top-quality luxury globes 'For State and Business Leaders' in the 1930s, converted to a decorative-modernist style for its post-war luxury globes. In contrast to the functionalist harshness of the American globes on plywood cradles or steel wires, these globes expressed a lush elegance. Columbus introduced the Duo-Leucht Globus, a double-image illuminated glass sphere, elegantly rotating on a curvy brass stand or a precious solid hardwood base. The cartography featured detailed geographical information, refined graphics, and a balanced color palette. Some models were custom-made, such as an exquisite postwar floor globe by Columbus with a stand shaped like a mahogany wishbone carrying a large, illuminated sphere. It stood in the office of the first West German chancellor, Konrad Adenauer (1876-1967) [no. 139].

Diplomat and Aristocrat

Ernst Kremling (1901-1977) founded the Munich road map company JRO Verlag in the 1920s. After the war, JRO produced luxury globes designed in an elegant mid-century style, using high-quality materials [no. 150]. The company even made motorized globes up to one hundred twenty-eight centimeters in diameter, such as a custom-made globe for the Pope [see page 94]. American globe makers also featured luxury globes in their catalogs, as highlighted in Replogle's advertising: *"Brand new base of solid genuine American walnut, rich gunstock finish, has smart sculptured styling that will enhance your finest room. - Far more beautiful!"* For those with traditional tastes, spheres and stands in eclectic historical styles remained available. Globe names such as the Diplomat by Replogle, or the Aristocrat by Weber Costello left no doubt about their intended audience.

What the gentleman prefers for Christmas
Eames Lounge Chair / *Playboy Magazine*, November 1958

32
Flemmings Verlag was a well-known German map publishing house in Hamburg, founded by Carl Flemming (1806-1878). The company started selling luxury globes after the Second World War. This model from the early 1960s features a wavy base of precious wood, shaped like the neck of a violin. In a spectacular balancing act, a metal ring supports the thirty-six-centimeter sphere and the metal meridian. The map gores display a refined cartographic design, with warm-colored landmasses contrasting against turquoise oceans. The strong shellac coating gives this classy object a glossy shine.

PSYCHEDELIC
GLOBES

Fluid Patterns

Danish architect Verner Panton (1926-1998) created the revolutionary Panton Chair in 1967. It was the first monolithic plastic-molded chair, available in many vivid colors. In a single gesture, a smooth cantilevered construction flows from top to bottom, forming a fluid S-shape with rounded-off edges. The Panton Chair became an icon of the psychedelic pop culture of the swinging sixties. The postwar generation of baby boomers challenged their parents' social conventions and experimented with sex, drugs, and rock music. They created a new psychedelic style, a departure from harsh modernism, using fluid hallucinatory patterns and bright colors. The joyful style found its way into graphic design, such as on Long-Play record covers, in psychedelic music, exuberant flowery fashion, funky interiors, and even some globes.

Good Vibrations

The use of bright-colored plastic enabled designers to capture the Zeitgeist. Plastic has significant advantages as the basic material for furniture and utensils. It is lightweight yet strong, and it can be cast into all kinds of curvy shapes and cheerful colors. Many plastic objects became icons of popular culture in an optimistic era, from Lego blocks to Olivetti typewriters. Globe makers also abandoned the seriousness of their traditional globes and created playful globes on plastic stands. Scan-Globe, the Danish globe maker, was obviously inspired by contemporary Danish furniture and psychedelic design. In the 1970s, they created lithographed metal spheres of Moon and Mars globes with airbrushed maps that resemble the dazzling liquid slides used in lightshows of that period. These were mounted on top of brightly colored plastic mono-legged stands [nos. 43, 45].

Wishbone Stand

One remarkable globe design features a tin sphere suspended in a plastic wishbone stand [no. 44]. The plastic full meridian creates a gimbal suspension, allowing the globe to swivel in all directions. The wishbone comes in bright green, red, or blue and is available as both a terrestrial and a lunar model. The globe does not have a brand logo and is available in several languages, suggesting it is manufactured as a nameless product for export. It features an identical lithographed tin sphere as some globes by the German company MS-Toys, though this model does not appear in their catalog. Various MS-Toys models have a mono-sculptural plastic stand in bright colors. In one of their models, the base and the meridian merge together in a single cantilevered S-shape that winds its way up, possibly inspired by the Panton Chair.

Panton Chair
Designer
Verner
Panton /
Denmark,
1967

33
J. Chein in New Jersey created this Moon globe during the 1960s space age. The white plastic monolith stand has a circular base that morphs from a single leg into three fingers with magnetic tips, supporting the nine-inch tin sphere. It is lithographed with a fuzzy-blueish photographic map. The names of the craters are speckled in electric green, creating a mysterious psychedelic effect. Early versions of this model showed a blank far side of the Moon, but this one, dating before the Apollo 11 moon landing, already includes the far side.

More Art Deco Globes

34

Paluzie in Barcelona created elegant art deco-styled globe models with sophisticated graphic design and high-quality lithography. This twenty-five-centimeter illuminated glass sphere rests on a circular-stepped metal base. The cartography is based on border coloring, allowing the landmasses to be depicted in a pale-beige color.

35

Paris-based Girard, Barrère & Thomas created a range of art deco-style globes. This compact 1940s model features a sixteen-centimeter illuminated glass sphere covered with paper map gores. The cartography is well balanced, with vivid colors and swirling sea currents. The base is a simple aluminum ring that reinforces the abstract composition.

36

Columbus Verlag in Berlin custom-made this decorative globe for the cutlery factory WMF in the 1930s. The eye catcher is the gold-painted Japanese tree that creeps up the greenish metal stand. Atop the stand is a twenty-centimeter sphere with paper gores in matching colors, which shows a post-First World War cartography.

More Streamline Globes

37

British map and globe maker George Philip & Son distributed this small six-inch terrestrial globe in the 1930s. The sphere, with its yellowish paper gores, rests directly onto the triangular cast-iron base painted in a metallic-bronze color. The composition of a circle on a triangle enhances the impression of the inclined planet floating in space.

38

Replogle from Chicago created various streamline-styled metal globe stands in the 1930s. This ten-inch model features an elaborate cast-aluminum base with a graded meridian rising up. The composition suggests speed: a world cutting through clouds, perhaps a reference to the winged feet of Mercury, the Greek god of speed.

39

Chicago globe maker Weber Costello also produced several streamline-styled globe stands in the 1930s. This model boasts a shiny cast-metal stand, shaped like an airplane. By today's standards, it teeters on the edge of kitsch, but it was fitting for its time. The black ocean map of the ten-inch sphere blends nicely with its shiny base.

More Wire Stand Globes

40

Chicago globe maker Replogle created globes with steel wire stands in the 1950s. This particular model, an eight-inch lithographed tin sphere called the Simplified Globe has a modernist design featuring playful details such as a winged nut to fix the metal meridian, a rack of braided steel wire reminiscent of kitchenware, and red plastic toes.

41

Perrina in Paris made elegant globes in the 1950s. This twenty-one-centimeter illuminated glass sphere is covered with paper map gores. The cartography is refined, with warm colors that blend nicely when the globe is lit. The stand consists of a delicate metal wire sledge and a bent double wire meridian, which holds the North and South Poles.

42

In the 1950s, European globe makers released various models supported by steel wire tripod stands, such as this joyful globe by Girard, Barrère & Thomas in Paris. The fifteen-centimeter pasteboard sphere, with its fresh-colored paper gores, floats on a steel wire base and meridian. The composition expresses the optimistic mood of the postwar period.

More Psychedelic Globes

43

Scan-Globe, the Danish subsidiary of Replogle, was clearly inspired by contemporary Danish furniture when it made this fifteen-centimeter Moon globe in the early 1970s. It presents a lithographed metal sphere with a detailed lunar map. In contrast to heavy traditional globes, it rests on a brightly colored mono-legged plastic stand.

44

This globe lacks a logo but was probably made by MS-Toys in Germany in the 1960s. The sixteen-centimeter lithographed tin sphere features a lunar map with craters and landing sites. It hinges on a plastic wishbone stand, which can swivel in all directions. It was available in green, red or blue, and also as a terrestrial globe.

45

Scan-Globe in Copenhagen created this fifteen-centimeter Mars globe in the 1970s. Its speculative map was airbrushed by Replogle's chief cartographer LeRoy M. Tolman, based on the low-resolution photographs of the planet taken by Mariner 4. The lithographed metal sphere resembles a liquid slide show, perched on a brightly colored plastic stand.

EDU-
CA-
CA-
TION

One World

Education was one of the main factors of success in the twentieth century. It elevated people from poverty, helped children to understand the complexity of the world they live in, and build a better society.

Globes played an important role in education as visual tools. A stunning example was the transparent globe developed by Robert H. Farquhar in Philadelphia during the late 1940s. He manufactured Plexiglas globes, on which he serigraphed the coastlines and borders, the longitudes and latitudes, all in vivid colors. Students could take the globe apart and use the two separate hemispheres for thematic hands-on projects. Farquhar called his educational ambition 'One World'. The idea was that looking through a transparent globe, one could more clearly see the relation between people and places on opposite sides of the world. He believed that it would lead to a better understanding of each other. Farquhar's transparent globes were excellent teaching tools, and many American schools purchased copies.

This promotional picture from the 1950s was accompanied by an explanation of the didactic approach: *"Marking crayons are provided so that any type of information may be put on the globe, such as, air currents, ocean currents, great circle routes, explorer routes, etc. The crayon can be easily wiped off with a soft cloth."*

Education of the masses became one of the priorities in the twentieth century. The globe emerged as a popular medium for imparting knowledge, serving as an educational tool in schools and a training instrument for the military. Thematic globes illustrated the latest scientific discoveries in physics, meteorology, or geology. Other globes became interactive learning tools, encouraging students to touch and write on them.

SCHOOL

GLOBES

Education for All

Italian physician Maria Montessori (1870-1952) developed an education approach for children at the beginning of the twentieth century. Her ideas remain a popular learning method today. She encouraged children to perform individual activities in the classroom and learn by discovery. Montessori designed learning tools that trigger experiences by appealing to the children's senses, such as a small tactile globe that shows the continents as colorful patches. In the period before Montessori, a majority of the European population was still illiterate, while at the same time, transforming an agricultural society into an industrial one required higher levels of education. Many countries adopted compulsory education laws, erected new schools, and standardized curricula. This led to the introduction of new subjects, such as geography, and a vast production of textbooks and teaching tools.

***Putman School
Sixth-Grade Class***
On every desk a
cheap six-inch
globe / *Boston,
1900s*

School Suppliers

A number of textbook publishers entered the new school market and transformed themselves into school supply companies. At the time, a cheap technology to print color images in textbooks was still lacking, so publishers offered complementary colored wall charts to hang in classrooms, depicting educational subjects. In most countries, the ministry of education made charts, maps and globes a mandatory part of school inventories. School suppliers hired cartographers to create maps and globes. As a result, one could find a globe in every classroom at the beginning of the twentieth century. It could be a large floor globe, where children stood around, or a suspended globe fixed to the ceiling, which the teacher could manipulate with cords and pulleys. Another option was cheap miniature globes, one for each child's desk [see picture above].

Simplified Cartography

School supplies could only be sold with accreditation from the ministry of education. For that reason, educational publishers had a head start and pushed traditional globe makers out of the market. Many of them became the leading globe producers in early-twentieth-century Europe, such as Paravia and Vallardi in Italy, Dalmau and Paluzie in Spain, Perthes and Reimer in Germany, or Forest in France. In the U.S., new school supply companies produced globes with simplified cartography, adapted to the level of the pupils. Examples include Hammond, Nystrom, Weber Costello, and Denoyer-Geppert. In the Netherlands, J.B. Wolters, publisher of the standard school atlas, made a school globe in the early twentieth century, edited by G.L. van Balen (1870-1943). The small paper sphere spins on a sturdy tin stand, making it easy to carry around [no. 57].

46
Weber Costello Company in Chicago patented this School Globe in 1909 and sold it until the Second World War. The so-called Peerless six-inch globe is made from a pasteboard sphere with paper map gores and fixed to a simple steel wire base. The wire stand became a solution that several school globe makers used in the first half of the twentieth century to reduce production costs. The idea was to provide a cheap globe on the desk of every student, as an alternative to a single large globe in front of the classroom. Despite its small size, the cartography is very detailed.

BLACK SLATE

Chalk-and-Talk

Scottish education reformer James Pillans (1778-1864) hung a large slate plate on the wall of his classroom to write on. His ground-breaking invention, the blackboard, was recognized as a classroom necessity by the British Council of Education as early as 1844. Chalk-and-Talk became the standard teaching method: the teacher addresses the students while visualizing the subject on the blackboard. The spherical blackboard followed in 1856 when German globe maker Joseph A. Brandegger (1797-1890) invented the Induktionsglobus, which allowed ideas to be induced onto the empty sphere. A year later, Forrest Shepherd (1800-1888) patented a black slate globe in the U.S., coated with a black substance made by the Candee Rubber Company in New Haven, CT. Another name for it is blank slate globe, referring to its lack of cartography.

Wipe-Off

The most extreme black slate globes are pitch black without geographical indications. Some have a grid of latitudes and longitudes, while others show the contours of landmasses. Most are mounted directly on a pin to avoid a meridian, making the globe easily accessible for drawing. In the postwar period, black slate globes were popular hands-on teaching tools: students could sketch countries, rivers, or time zones on them or draw principles of astronomy or goniometry. After the lesson, the chalk could easily be wiped off to empty the globe for the next assignment. Otto E. Geppert (1890-1970) of the Denoyer-Geppert Company wrote: *"Sometimes I refer to this globe as similar to a storage battery. Looking at a storage battery we do not see very much, but it has a remarkable potential. It starts the car motor, it lights the lights, it plays the radio".*

Squeaking Felt Pens

Throughout the twentieth century, many globe makers promoted black slate globes. In Europe, this included Dietrich Reimer and Paul Räth, while in the U.S. companies like Cram, Denoyer-Geppert, Rand McNally, and Nystrom were notable producers [nos. 63-65]. Weber Costello, the largest U.S. producer of chalk sticks and wipers, made many black slate globes, including an elegant art nouveau model on a cast-iron stand. After Sidney Rosenthal (1907-1979) invented the Magic Marker in the 1950s, the powder dust and scratching sound of chalk on a blackboard were gradually replaced by the squeaking sound of an alcohol felt pen on whiteboard. The Danish Scan-Globe company published a whiteboard globe in the 1970s, made of white plastic, with coastlines and borders in dotted black lines. One could draw on its smooth surface with felt pens and wipe it off afterward with a cloth [no. 62].

Navigation training
U.S. Navy Quartermasters working on a blank slate globe / *Norfolk, Virginia, 1944*

47
A Czechoslovakian school supplier manufactured this black slate globe in the 1970s. The model has a thirty-centimeter black sphere that features only a few basic geographical indications, such as the shape of the coastlines as red lines, the equator and the Greenwich meridian as yellow lines, and the Tropic of Cancer and Capricorn as blue lines. It provides teachers and students with a reference for drawing geographical information with chalk on this three-dimensional map. The black sphere is mounted on a graded meridian of translucent plastic and attached to a modernist steel wire stand.

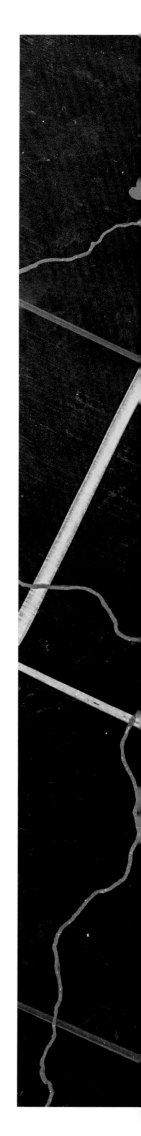

PROJECT PROBLEM

GLOBES

Pilot Training

American test pilot Chuck Jaeger (1923-1920) was the first to break the sound barrier. In his experimental X-1 rocket plane, he flew at 1,126 km/hour in 1947. That same year, the U.S. Air Force became a separate branch of the U.S. Armed Forces, highlighting the crucial importance of military aviation, proven during the Second World War. A new training tool for pilots was the Project Globe, also known as the Aviation Globe, which became popular at U.S. military academies. Its large steel sphere is perfect for plotting flight routes or drawing military strategies. On a flat map, a straight flight route between two airports is distorted into a curved line due to the projection of a round sphere onto a flat surface. However, when tracing a flight route on a globe, one can easily see that the shortest connections are straight lines that curve over the globe.

Problems and Activities

In the upcoming air age, it was assumed that everybody would use airplanes or have a job linked to aviation. Therefore, the U.S. Civil Aeronautics Administration recommended integrating pre-flights aeronautics into the high school program in 1943. It included navigation, meteorology, aerodynamics, geography and more. Schools were advised to purchase a Project-Problem Globe or Activity Globe from suppliers such as Denoyer-Geppert or Nystrom. The models feature a large twenty-one to twenty-four-inch diameter sphere made of spun metal, allowing a group of students to work on it together. Fixed to a heavy-duty metal stand, the globe is sturdy enough to withstand student activity. Its cartography is simple, with landmasses shown as monochrome surfaces in black, green or orange, contrasting with the black or blue oceans [no. 179].

Didactical Plasticine

British art teacher William Harbutt (1844-1921) invented Plasticine in 1897, a non-drying modeling clay he used in his sculpture classes to help stimulate students' free expression. It became a popular educational material, and half a century later, Plasticine was used as a didactic tool to teach geography on project globes. It perfectly blended with the new pedagogical methods of that time, which emphasized hands-on activities and group work. Cram's 1950s catalogs explained its function as follows: *"This Project Globe is used to show any information desired - to bring out physical, political, regional, historical aspects, etc. Modeling-clay can be used to build physical features. Your teaching needs will suggest many other uses."* After the plasticine is pasted on every corner of the world, it can be removed and the material can be reused for the next assignment.

Plasticine Mountains
Schoolgirl working on a Denoyer-Geppert Project Globe / *Chicago, 1950s*

48
Nystrom from Chicago made this model called a Project-Problem Globe from the 1940s until the 1980s. It could serve for geography lessons at schools as well as for pilot training at military institutes. The large twenty-four-inch sphere is made of spun aluminum, featuring a rudimentary map with black landmasses and blue oceans, separated by a golden coastline. The sphere hinges in a metal meridian, and a heavy disc base keeps the floor globe upright. The catalog states: *"color crayons may be used, the marks can be erased with a damp cloth, or better still, with a dry piece of chamois skin."*

BRAILLE
GLOBES

A World at your Fingertips

Louis Braille (1809-1852) was blinded as a young boy when he accidentally stuck an awl in his eye at his father's leather workshop. As a blind person, he made a monumental contribution to the emancipation of the visually impaired. He created a tactile reading system based on an earlier military system used by Napoleon's army that enabled officers to read messages in the dark. In 1829, Braille published his first work in relief script. Braille script would become the world standard and consisted of patterns of dots embossed in paper. By touching the pattern, one can read the text with one's fingertips. During the nineteenth century, schools for the blind opened in Europe, requiring special learning tools such as Braille script textbooks, tactile maps, and even Braille globes for geography lessons, to discover the world by touching the relief of mountains and the course of the rivers.

Embossed Cartography

Geography teacher Johann August Zeune (1778-1853) founded a School for the Blind in Berlin. As early as the 1820s, he developed raised relief globes for his students. Many globe makers followed his idea with various models of relief globes. In Australia during the 1920s, the *Blind, Deaf and Dumb Children Instruction Act* was passed, making the education of impaired children compulsory. Richard F. Tunley (1878-1968), an inventor and manufacturer of educational tools for the blind, was central to its passing. A year later, he created a handmade Braille globe consisting of a wooden sphere with metal plates screwed on in the shape of the landmasses and topographical names in embossed metal Braille script. Although initially developed for the visually impaired, relief globes were also offered to the general public, to visually illustrate the mountain ranges on Earth [nos. 21, 22, 23, 51, 145].

Fiberglass Mountains

Another blind man, Sherrod Dempsey (1828-1879), founded the American Printing House for the Blind in Louisville, Kentucky, in 1858. This non-profit organization, still active today, published many Braille script books and relief globes during the twentieth century. A fine example is a large thirty-two-inch floor globe designed in the 1950s by the Panoramic Studios in Philadelphia. Like most postwar Braille globes, it is made from fiberglass cast in a mold. Landmasses, mountains and rivers, longitudes, and latitudes are all in raised relief, allowing the blind to explore the geography by touch. The large size allows for a lot of detail, while the contrasting colors help the visually impaired. In Europe, Jörn Seebert made various Braille globes in the 1950s, such as the large fifty-centimeter handmade polyester resin model, with a fifty-to-one exaggeration of the heights [no. 22].

Tangible Geography
Blind children studying a large relief globe / U.S., 1914

49
The American Printing House for the Blind, in Louisville, Kentucky, distributed this Braille globe in the 1960s. It not only features landmasses, main rivers, and mountain ranges in raised relief, but also has the equator and the Tropic of Cancer and Capricorn as dotted lines. Large cities are marked with a single dot. Topographical names are cast in the twelve-inch plastic sphere using raised Braille script dots. The sphere rests loosely on a steel wire stand, allowing blind students to pick it up and examine the world from all sides by holding it in their hands and feeling the geography with their fingertips.

CRADLE

Rollglobus und Erdmesser

Austrian engineer and businessman Robert Haardt (1884-1962) was a globe collector and founder of the Globe Museum in Vienna. In 1935, he revolutionized globe making by the invention of the Rollglobus. In this model, the sphere is not fixed to a meridian, but rests freely on a ring [see opposite page]. The Rollglobus served both scientific and educational purposes: on a traditional globe, it is difficult to observe the southern hemisphere and the polar regions, as they are obstructed by the meridian or the horizon ring. A Rollglobus, however, has no top or bottom, just like Earth floating in space. One can pick up the sphere, view all sides, and hold it in one's hand for examination. An additional feature was the Erdmesser: a metric scale on the ring's base to measure distances between points on Earth by simply rolling the globe into the right position.

Ermesse die Welt!
Inventor
Robert Haardt
demonstrates his
Schüler-Rollglobus /
Vienna, 1930s

Tuffy Orb

In the U.S., the Rollglobus was called a cradle globe. Stripped of fragile elements, a cradle globe was a robust model for schools. Most American globe makers built cradle globes in the postwar era. Cram made a popular sixteen-inch school globe that rested on a metal ring. Cram's catalog echoes the words of Robert Haardt as it praises the cradle globe's educational virtues, *"... since the ball is not attached to the base, every student may hold the world in its lap. With his imagination thus stimulated, it is easy to visualize that the central point of his world is where he lives."* The tough plastic sphere, called Tuffy Orb, was advertised as unbreakable, not unimportant if a student accidentally dropped it on the floor. Cradle globes were offered with single and double rings, called Clear View Mounting [no. 179] or Horizon Ring Cradle Mounting [nos. 165, 178].

Geo-meter

Denoyer-Geppert created many cradle globes with a simple wooden cross-shaped base, such as the sixteen-inch Cartocraft Physical-Political Globe, which rests freely on a cradle, felt-lined for smooth movement and protection of the paper map gores [nos. 163, 178, 184]. Even a celestial model was available. The National Geographic Society ordered sophisticated cradle globes from Replogle in the 1960s to offer to the readers of its magazine. In this case, the twelve-inch globe rests on a transparent Lucite cradle base. It is equipped with a transparent horizon ring to calculate time and distance or to plot the orbit of satellites. The model came with a loose hemispherical transparent dome, the so-called Geo-meter, to measure Earth's surfaces. The cradle base remained popular for a long time, but the Geo-meter never caught on [no. 174].

50
Columbus Verlag produced several versions of Haardt's Rollglobus. This twelve-centimeter student model from the 1950s has a graded Bakelite scale ring for measuring distances. In 1936, Haardt's praised his globe as an object of hope: *"The globe is most necessary for the youth, who today, due to the great development of radio and aviation, is much more preoccupied with universal problems than a few decades ago. It is a great advantage for the student if he can acquire knowledge about the world in a playful way. It anchors in him an element that unites peoples and maintains peace".* Haardt's hope was in vain: two years later, Hitler triumphantly marched into Vienna.

"The globe is most necessary for the youth, who today, due to the great development of radio and aviation, is much more preoccupied with universal problems than a few decades ago. It is a great advantage for the student if he can acquire knowledge about the world in a playful way. It anchors in him an element that unites peoples and maintains peace".

Members of Wembley County School's Astronomical Society gather around their suspended black slate globe during one of the evening classes / *London, 1957*

Robert Haardt, 1936
Inventor of the Rollglobus
Founder of the Vienna Globe Museum

PHYSICAL

GLOBES

Race to the Poles

American explorer Robert Peary (1856-1920) reached the North Pole in 1909, only two years before his Norwegian rival Roald Amundsen (1872-1928) reached the South Pole. At the beginning of the twentieth century, humanity had visited the most remote places on Earth, and the map of the world was complete. A mayor issue for cartographers was how to graphically represent the natural elements on a globe. A 'political globe' depicts the man-made world, such as countries, borders, or time zones, while a 'physical globe' depict the natural world, including mountains, rivers, and sea currents. Today, we understand that a black dashed line represents a railroad, while a winding single line represents a river. Without even thinking about it, we see blue as water, brown as mountains, and green as plains. However, in the early twentieth century, these conventions were still being established.

Political or Physical

Hermann Haack (1872-1966) set a new standard for the cartography of globes in the 1910s. As head of cartography at the German publishing house Perthes, he created very detailed, high-quality map gores. He used eight color shades for land heights, ranging from light green to yellow, and orange to dark brown, and five shades of blue for sea depths, and rationalized the line types and letter fonts. Haack solved the graphic problem of competing political and physical information by splitting it into two globes: a 'political' and a 'physical' one. It was a great idea, but a commercial failure, as school budgets did not allow for the purchase of two globes. To his disappointment, Haack had to merge the two types again. In an effort to present it as a new invention, he called it the Physisch-Politische Einheitsglobus (Physical-Political Unified Globe).

Double Image

Paul Oestergaard Jr. (1904-1975) was the second generation at Columbus Verlag, the Stuttgart globe maker. He solved the problem of separating political and physical information on a single globe in the 1950s when he patented the Duo Light Globe, also known as the Double Image Globe. The globe consists of an illuminated glass sphere covered with map gores. When the lamp is off, the observer sees a political globe, but when the light is switched on, the physical globe appears. The secret of the changing image lies in the printing: the political map is printed on the outside of the paper, while the physical map is printed on the inside. When the light is on, the physical map overrides the political map. This technology was very successful and became the world standard for combined political-physical globes in the second half of the twentieth century [nos. 139, 149, 152].

The Three Polar Stars
Explorers Roald Amundsen, Ernest Shackleton and Robert Peary / *Philadelphia, 1913*

51
Räthgloben in Leipzig was the main East German globe maker during the Cold War. This striking model is a physical globe from the 1950s. The thirty-three-centimeter papier-mâché sphere features paper map gores that depict height differences through an exaggerated 3D relief and a color system called Farbenplastik, which pairs heights to colors: green to yellow for lowlands, orange to brown for mountains. The wooden meridian embraces the sphere and blends into a heavy base, creating an iconic art deco globe. The model was commercialized during the 1930s by Berlin globe maker Dietrich Reimer.

CLIMATE

The Weather

German explorer Alexander von Humboldt (1769-1859) was the greatest scientific voyager of his time. For centuries, European powers had undertaken expeditions with profit as their primary goal: finding trade routes and foreign markets. Geographical discovery was at most a byproduct of an economic quest. Von Humboldt broke with this tradition and undertook expeditions that focused strictly on scientific research and laid the foundation for new sciences, such as physical geography and meteorology. This shift moved the focus from topography to understanding the physical processes that shape the Earth, such as the weather, subterraneous layers, and the flow of air and water. During the twentieth century, globe makers created various thematic demonstration globes to disseminate this new scientific knowledge of the Earth's natural phenomena.

Passat and Monsuun

Carl Kassner (1864-1950), a professor at the Prusian Meteorological Institute, conceived a Meteorologischer Globus, produced in 1907 by globe maker Dietrich Reimer Verlag in Berlin. The drawings on the map gores feature high and low-pressure zones, temperatures, air currents, climate zones, the hot equator and the cold poles, the influence of landmasses, mountains, and sea currents on the climate, the trade winds (Passat) and the monsoon (Monsuun) and more. Kassner developed graphics to demonstrate various meteorological phenomena. He solved the problem of the seasonal differences by creating a pair of globes: a January globe and a July globe. An interesting feature was the foldable meridian. It allowed the user to turn the sphere upside down without removing it from its stand, so that the southern hemisphere could be observed without obstruction.

Arctic and Tropic

Georg Jensch (1908-1978), a Berlin professor of geography and cartography, made another attempt to create a meteorological globe in 1969. The model was produced by globe maker JRO in Munich and published by Kiepert Verlag in Berlin. Jensch argued that since weather phenomena on one side of the world, such as cold arctic winds or hot tropical sea streams, influence the climate on another side of the world, a globe was a much better instrument than a flat map to visualize the interaction of meteorological phenomena. For the coloring of the map, he followed the so-called Koppen-Geiger Climate Zone System. This is a classification of climate zones developed in the twentieth century, based on the local occurrence of plant types in combination with the levels of rainfall, temperature, and other factors, which provides a more balanced division of the various zones.

Dissecting the Earth
Employee coloring a large Geological Globe at the Geological Museum / *London, 1930s*

52

Munich Re Insurance Company published this thirty-centimeter Globe of Natural Hazards in 1998, printed by Mairs Geographischer Verlag. The map gores, compiled from the company's statistics, show the risk zones on Earth that are frequently hit by natural disasters. It includes phenomena such as tropical storms, cyclones, heatwaves, limits of pack ice, tsunamis and high sea waves, volcanic eruptions and earthquakes. Consequently, the globe's surface is covered with strong colors and pointy arrows, making it a terrifying conversation piece in the boardroom of any insurance company.

Globe of Natural Hazards

TECTONIC

GLOBES

Continents Adrift

German geophysicist and polar explorer Alfred Wegener (1880-1930) published his book *The Origin of Continents and Oceans* in 1915, in which he proposed the hypothesis that the continents had once been interconnected but had drifted apart over millions of years. The observation that the coastlines of the continents fit together like puzzle pieces was first made by Flemish cartographer Mercator in the sixteenth century. Wegener found additional evidence in the discovery of similar fossil plants and animals on both sides of the Atlantic Ocean. His Tectonic Plate Theory, as it was later called, was controversial: it took fifty years for the scientific community to accept it. For globe makers it was a complex task to represent a dynamic geological process happening on the ocean floor, involving time and transformation, on the static surface of a globe.

Lady and the Submarine

American oceanographer Mary Tharp (1920-2006) worked at the Lamont-Doherty Earth Observatory in New York in the 1950s. Following the social customs of the time, she was not allowed to join her male colleagues on a submarine research trip in the Atlantic Ocean. She had to stay behind and draw the map of the ocean floor based on the data sent by her male colleagues. The gender norms of the fifties became Tharp's path to fame: when she pieced together the ocean floor map, an amazing rift appeared in the middle of the Atlantic that allowed her to proof Wegener's Tectonic Plate Theory. Tharp not only drew a map, but even created a Bathymetric Globe, a reverse terrestrial globe that depicts the relief of the ocean floor, with its ridges and trenches in different colors, while the continental landmasses remain black, without information [see picture above].

Ocean Floor Relief

In the 1960s and 1970s, globe makers took Tharp's map as a basis for globes that create the illusion of ocean floor relief through color and shading. Rand McNally offered The Aqua-Sphere or The Oceanographer. Replogle published the Nations of the World with Sea Relief, and the World Ocean Series. German cartographer Kurt Ziesing created a tectonic globe, published by VEB Hermann Haack, which shows color-coded tectonic plates and red faults lines. Robert H. Farquhar created a transparent globe that shows the fault lines continuing to the other side [no. 8]. A clever solution is the Drift Globe by a small workshop in Winthrop, U.S. Plastic shapes, attached to the sphere with Velcro tape, represent the tectonic plates. Users can take them on and off and move them to different positions, introducing time as the fourth dimension of a globe [no. 65].

Bathymetric Globe and Ocean Floor Map
Marie Tharp standing at her drawing board / *New York, 1950s*

53
The National Geographical Society in Washington D.C. produced this twelve-inch globe called the Physical Globe in the 1970s. It was drawn by William H. Bond (1897-1980) of the National Geographic Map Division, based on Marie Tharp's ocean floor map. The globe has an overall greenish color that diminishes the difference between land and sea. It suggests a world without water, where one can observe the relief of the Earth's crust as a continuous surface, a 3D illusion generated by a shading technique. The tall cradle stand of smoked Lucite reflects the design trends of the 1970s.

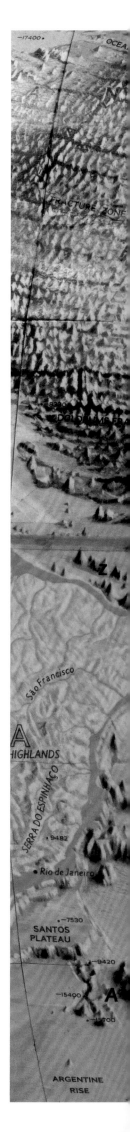

SUBTERRANEOUS

GLOBES

Hollow Earth

Jules Verne's (1928-1905) book *Journey to the Center of the Earth* introduced the idea that an unknown civilization lives inside the Earth. Subterranean fiction blossomed not only as a literary genre in the early twentieth century, but also as part of religious scientific theories. Cyrus Teed (1839-1908), an American physician, founded a utopian community in New York to promote Koreshanity, the belief that we live on the convex inner surface of a hollow Earth, looking towards the universe in its interior. Teed made a hollow globe to visualize this concept. Another hollow Earth theory was proposed by Marshall B. Gardner (1854-1937). In his book *A Journey to the Earth's Interior*, he claimed that there are large openings at the poles to enter the hollow Earth, which has a sun inside. Gardner patented a design for a hollow globe to demonstrate his theory.

Hollow Earth Globe
Demonstration
by Hedwig Michel
(1892-1982),
President of the
Koreshan Unity /
Florida, 1961

Journey to the Center of the Earth

The scientific understanding of the subterraneous Earth was unraveled through the combination of geological research and Charles Darwin's (1809-1882) theory of evolution. By identifying the fossils found in each geological layer, it is possible to construct a timeline of the Earth's stratification. These geophysical findings created yet another challenge for globe makers: how to represent underground layers on a globe? German paleontologist Wilhelm Dames (1843-1898) was the first to create a geological globe, produced by Dietrich Reimer around 1900. He used sixteen colors to represent the different geological time scales within the subterranean Earth. A hundred years later, the International Commission for the Geological Map of the World compiled the Earth Geological Globe, by digitizing the most recent geological data and transferring them as a color code on the globe [no. 197].

Layers and Cores

A clever solution for a subterranean globe was proposed by Mátyás Márton and Balázs Kovács in Hungary, published by Kártográfia Vállalat in the 1980s. They addressed the challenge by cutting the globe into segments. The model can be taken apart to reveal the deep Earth layers, all the way to the core. On the outside, geological formations, plains and fault lines are depicted, while the inside displays various underground layers. Color codes illustrate the different geophysical characteristics of the Earth. A similar globe for teaching purposes was manufactured in Czechoslovakia in the 1970s. On the outside, the sphere only shows oceans in blue and landmasses in black, with red grade lines on top. A segment of the sphere can be removed to expose the subterranean layers in vivid yellow and orange colors. It is a true educational demonstration tool [no. 66].

54

Andrea Grassi, an amateur globe maker from Italy, handmade this twenty-centimeter demonstration model of the Sun in the 2000s. For obvious reasons, not many solar globes have been built: the Sun is a giant fire ball without a representable surface. Grassi recreated the idea of the burning flames on the outside, while opening up the inside of the globe to reveal the various layers of the core. The bright-yellow and orange colors suggest the raging hot fire. This unique Sun globe serves an educational purpose but is also an attractive decorative object.

DAY-AND-NIGHT

GLOBES

Circadian Rhythm

Austrian neurologist Sigmund Freud (1856-1939) pioneered psychoanalysis, the theory of the unconscious, which dominated the twentieth century. In his book *Die Traumdeutung*, he introduced the interpretation of dreams. Dreams occur during our sleep-wake cycle, which is controlled by an internal biological clock that follows the twenty-four-hour cycle of night and day on Earth, a process called the Circadian rhythm. In the twentieth century, the Circadian rhythm became even more apparent when it was disrupted by long-distance air travel, leading to a novel condition known as jet lag. The day-and-night cycle is the result of the Earth revolving around the Sun while spinning on its axis, as proven five hundred years ago by Polish astronomer Nicolaus Copernicus (1473-1543). Various astronomical instruments have been created since then to demonstrate the day-and-night phenomenon.

Trippensee Tellurian

Canadian Alexander Laing (1845-1917) patented a mechanical Tellurian in 1896. A Tellurian is a model of the Earth, the Sun, and the Moon with movable parts, connected with pulleys and chords, to demonstrate orbits and rotations, the day-and-night cycle, the four seasons, and the phases of the Moon. At the position of the Sun, there is an electric light to simulate sunlight. Ten years later, the Trippensee Brothers, a company supplying automobile parts for Buick and Ford in Detroit, began producing Laing's Tellurian and improved the design using chains and gears. The early Trippensee Tellurians were built of wood and brass, while the later ones were made of Bakelite and plastic. At the end of the twentieth century, the company was acquired by Science First, a scientific learning aids producer in Buffalo, which still produces the Trippensee Tellurian today [no. 108].

Night Caps

The George F. Cram Company from Indianapolis started producing the Sun Ray Globe in the 1930s. This terrestrial globe is equipped with an automatic sun ray and season indicator and a day-and-night meridian. The meridians can be adjusted and pointed to a specific time of the year, printed on its circular base. It demonstrates astronomical phenomena such as the seasons, day and night, the position of the Sun in relation to the Earth, the angle of the sun rays, and more [no. 67]. A straightforward solution to show the day-and-night cycle is the Capped Globe. This type of globe has a rotating shield that covers the part of the globe where it is nighttime. One example is the Kugelhauben-Globus patented around 1910 by Johann Georg Rothaug (1850-1924) in Vienna. Another example is Dietrich Reimer's 1930 clock globe [no. 78].

The difference between day and night
Eveready Flashlights advertisement, painting by Frances Tipton Hunter / U.S., 1930s

55

Peter Oestergaard of Columbus Verlag in Stuttgart, renowned for its innovations, modernized the concept of the capped globe and patented a special day-and-night globe in the 1973, known as the Planet Erde Globe. The thirty-centimeter illuminated translucent plastic sphere contains a cap inside, resulting in illuminated and darkened areas that simulate day and night. As the globe rotates, the light zone progresses, akin to reality. By adjusting the meridian, users can shift the Earth's inclination to demonstrate the transition of the Sun between the northern and southern hemispheres during various seasons of the year.

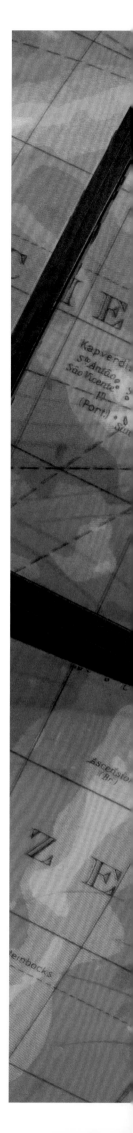

More Metal School Globes

56

Around 1900, German tin toy factories began manufacturing and exporting miniature tin globes. This model features a three-inch sphere, crafted from two tin hemispheres, with a simplified lithographed cartography due to its small size. To maintain affordability, a single steel wire is bent to form a simple stand. An ideal tool for teaching children geography.

57

J.B. Wolters, Dutch publisher of school books and atlases, created this school globe in the 1910s. The miniature ten-centimeter sphere incorporates paper map gores and is suspended within a four-legged tin stool with a graded horizon ring. Its compact size and tin construction make it an affordable and robust portable globe for students.

58

Denoyer Geppert, a Chicago-based globe maker, introduced this school globe in the 1930s. The model, called the All-Metal Pupils Globe, was a compact four-inch globe priced at just 50 cents, making it accessible to all. Marketed as an alternative to paper globes, it boasted an indestructible material. The base includes geographical explanations.

More Paper School Globes

59

Vallardi, an Italian publisher of educational materials in Milan, produced small school globes. This model from the 1930s has a fifteen-centimeter sphere with paper map gores featuring cartography by A. Minelli. Unlike Vallardi's larger models, these school globes lack a meridian. They are mounted directly on the South Pole and connected to a wooden tripod.

60

Vallardi evolved its small school globes over time. This 1940s model is smaller, with a ten-centimeter sphere. The map has been redrawn, and the wooden stand now has an asymmetrical shape, with a triangular base and a leg mirroring the inclination of the Earth. Later, the wooden leg would eventually transform into a dynamic curved Bakelite fin.

61

The Soviet Union produced its own school globes. This model dates from the 1980s. Although it is a full plastic model, it still has paper gores, glued on the fifteen-centimeter sphere. The cartographic image shows a physical map. It is equipped with a metal meridian and a transparent plastic graded arc for latitude measurement.

More Blank Slate Globes

62

In the 1970s, the Danish Scan-Globe introduced this induction globe. The thirty-centimeter sphere, made of cream-colored plastic, features only dotted lines to mark coastlines and borders. Students can use felt pens to draw on it and wipe off their markings afterwards with a cloth. The globe is mounted on a circular plastic base with a metal meridian.

63

George F. Cram, the Indianapolis-based globe maker, created this twelve-inch blank slate globe in the 1930s. Teachers and student use chalk to draw geographical notations on its empty sphere, akin to a 3D blackboard. The surface has a pleasing greenish color. It is mounted on a bronze meridian resting atop a sturdy metal base.

64

This twelve-inch globe was manufactured by British map makers W. & A.K. Johnston & G.W. Bacon. The model was already produced around 1900 and was sold until the 1950s. In this version, it is supported by a wooden cross base without meridian, allowing all surfaces to be easily reached for marking and scribbling with chalk on the black-varnished paper.

More Scientific Globes

65

Drift Globe, based in Winthrop, WA, U.S. manufactured this tectonic globe in the 1980s. It comprises a black twelve-inch plastic sphere resting on a wooden base. Loose plastic shapes representing the continents are affixed to the sphere with Velcro tape. By removing and repositioning these shapes, users can visualize the theory of continental drift.

66

This educational globe was made in Czechoslovakia in the 1970s. The surface of the thirty-three-centimeter plastic sphere displays blue landmasses against a black backdrop of oceans, accented with red grade lines. A segment can be removed to reveal the hot subterraneous layers of the Earth's interior, depicted on paper in vibrant red and yellow hues.

67

Cram from Indianapolis produced a twelve-inch Sun Ray Globe from the 1930s to the 1950s. It is equipped with an automatic sun ray and season indicator with a day-and-night meridian. When turned to the corresponding time of the year, as indicated on its base, the globe displays the angle of the Sun's rays and the boundary of daylight.

SO-CIETY

World Power

Traditionally, giant floor globes adorned the castles and palaces of powerful men. A monumental globe symbolized the worldly power of the owner, whether they were a monarch or even the representative of God on Earth, as in the case of Pope John XXIII (1881-1963).

Shortly after his accession in 1958, the new Pope commissioned a monumental custom-made world globe. The project was entrusted to Father Heinrich Emmerich (1901-1984) of the Society of the Divine Word, a congregation of missionaries. Emmerich, who had already created the *Atlas of the Catholic Missions* drew the map gores for the globe, which included two thousand two hundred dioceses and church provinces around the world, marked with special icons and colors. For each diocese, the name, location, dignity, local rite, and the Roman congregation on which it depended were mentioned. The story goes that when foreign dignitaries or heads of state came to visit the Vatican, the Pope always invited them to point out their hometown on his globe. This gave him the opportunity to demonstrate to his guests the power of Catholicism around the world.

The picture shows the Pope and his globe in the early 1960s. The 127-centimeter diameter globe was produced by JRO in Munich. Today, one can still find the Pope's handwriting on it: he made corrections, added unmarked dioceses, and pointed out holy places where he had been on pilgrimage.

In the twentieth century, numerous dramatic events unfolded. Globes were used to illustrate the most recent human activities in politics, war, and economics. They also became vehicles for showcasing the many inventions that emerged, including radios, telephones, and airplanes. The twentieth-century globe had to be versatile, adapting to the rapid developments in a globalizing society.

POLITICAL

Shattering Europe

Joseph Stalin (1878-1953), Winston Churchill (1874-1965) and Franklin D. Roosevelt (1882-1945) convened in Yalta, shortly before the end of the Second World War. The three Allied leaders redrew the map of Europe, dividing the continent into a Soviet zone and a Western zone. This redrawing altered European borders: Germany shrunk and split in two, Poland shifted westward. The European map had already undergone significant changes twenty years earlier as a result of the First World War, which saw the Austrian Empire break apart into countries such as Czechoslovakia, Hungary, and Yugoslavia. The European map splintered again in the 1990s, with twenty-one new countries emerging after the breakup of both the Soviet Union and Yugoslavia. The political changes of the twentieth century drove globe makers to despair, as they constantly had to redraw and reprint the map gores.

Decolonizing Africa

During the first half of the twentieth century, the map of Africa required only seven colors [no. 3]. They represented the European colonial powers, which, at the Berlin Colonial Conference of 1888, had divided the continent among themselves without inviting a single African leader. In the three decades following the Second World War, fifty-four African states regained independence, often after bloody conflicts. This turned the African map into an ever-changing patchwork of colors, making new globes quickly outdated and unsellable [no. 68]. In a 1970s newspaper interview, Replogle president William C. Nickels (1921-2002) complained: *"... Africa is a globe maker's nightmare ...."* For scholars, on the contrary, map changes are a handy tool for determining a globe's age: if a certain state is present, the globe dates back to no earlier than that state's independence.

Political Propaganda

The true purpose of the patchwork of states on a globe is to showcase geopolitical power, making a globe a perfect tool for propaganda. Iraqi President Saddam Hussein (1937-2006) ordered globes from the Italian globe maker Rico to promote Pan-Arabism, the ambition to unite all Arab countries, with the condition that all Arab countries would be identically colored in bright orange. Globe makers do not hesitate to accommodate different political realities for different export destinations. For example, Argentinian globes show the Falkland Islands as part of Argentina, Chinese globes depict Taiwan as part of China, and Indian globes include Kashmir as part of India. When asked about this practice, LeRoy M. Tolman (1931-2015), the chief cartographer of Replogle, responded phlegmatically: *"The customer is not always right. But he is always a customer".*

Invasion Maps and Atlases Here
Rand McNally's shop window during wartime / *Chicago, 1940s*

68
This mid-1970s political globe illustrates the decolonization of Africa. In the postwar period, fifty-four African countries gained independence, transforming the map of Africa into a vivid patchwork of contrasting colors. The twelve-inch Stereo Relief globe by Replogle features a raised relief surface. It sits on a modernist wooden stand, with an upright leg that fixes the meridian, offering a full view of the sphere. Replogle wrote: *"Specially designed for easy viewing ..."* and *"Brand-new base of solid, genuine American walnut, has smart sculptured styling that will enhance your finest room."*

WAR

War and Peace

British Prime Minister Neville Chamberlain (1869-1940) returned from Munich in 1938 with a supposed agreement between Nazi Germany, Italy, France and the U.K. At the airport, he waved a piece of paper, declaring: *"Peace for our Time!"* He was deceived by Hitler: a few months later, the Second World War broke out. From that moment, American globe sales dropped but surged again two years later when Japan attacked Pearl Harbor and the U.S. entered the war. During wartime, more globes are sold than during peacetime. People at home buy globes to localize the exotic cities, faraway beaches, and tropical jungles they hear about in the news, where their beloved sons and husbands are fighting. In a wartime advertisement for its globes, Replogle appealed to this human urge: *"Keeping up with the News is easy with a Replogle Globe."*

Vouchers and Coupons

Wartime is complicated for globe makers. Shifting frontlines can outdate a globe within days, creating unsellable stock. On the other hand, the instability discourages the public from buying a globe. To address this, Replogle included a coupon with their globes. If mailed back within 90 days after the signing of the peace treaty, the owner would receive updated map gores: *"... Do not hesitate to purchase this globe because it shows Europe as it was September 1, 1939. Boundary changes since then cannot be official or permanent because they must be ratified by a peace treaty ...."* Cram offered a Self-Revising Globe that guaranteed buyers a free map revision service after the war. *"The publishers of this Cram Globe will supply new, revised map sections accurately die cut and so prepared that with the aid of very simple instructions the owner may bring his globe up-to-date."*

Borders and Overlays

Replogle already struggled with the problem of border changes before the outbreak of the Second World War. In 1938, director Luther Replogle (1902-1981) suggested to his team to create a semi-transparent overlay sticker showing Nazi Germany's annexation of Austria and Sudetenland, which customers could affix to their existing globes. This idea sparked an internal dispute with chief cartographer Gustav Brueckmann, whose name featured on most Replogle globes. As a German immigrant, he refused to have his name on a globe that normalized aggression. In the 1990s, map problems became even more challenging when the Soviet Union and Yugoslavia imploded, and new countries emerged. By that time, most globes were made of plastic, making a map revision system impossible. Globe sales dropped, and many globe makers went out of business by the end of the century.

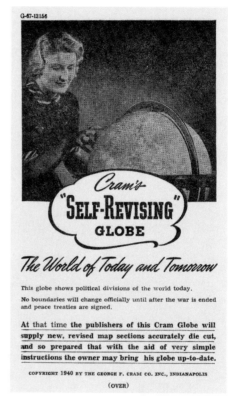

Self-Revising Globe
Cram's wartime voucher for globe revision /
U.S., 1940s

69
Weber Costello sold this twelve-inch globe in 1939, reflecting the imminent war with grim black landmasses contrasting against white oceans. The map is simplified, drawing attention to the thick blood-red political borders that would soon be fiercely contested. The aerial endeavors of the 1930 are depicted with dotted lines, showing the flight routes of Lindbergh, Hughes and Graf Zeppelin over the Atlantic, Byrd's North Pole crossing, and the China Clippers flying boats of the Pacific. The base is remarkable: in order to save metal for the war effort, the cast-metal airplane of earlier models was replaced by a wooden version.

ECONOMY

Knowledge is Power!

Belgian bibliographer Paul Otlet (1868-1944) had the ambition to collect all human knowledge. To achieve this, he pioneered a prototype internet based on index cards in the 1910s. Ultimately, it became a database of fifteen million typed cards. He commissioned modernist architect Le Corbusier (1887-1965) to design a world center of knowledge, the Mundaneum, in Geneva, but the project never materialized. In the same spirit, globes of that period were conceived as visual encyclopedias. Berlin globe maker Paul Oestergaard (1872-1956) proclaimed: "*Knowledge is Power! Knowledge of our Earth is of the utmost importance, especially in our busy time of steam and electricity, which spans all distances. Today, when the telegraph brings news from the furthest corners of the Earth in a very short time, it is essential to have a really good terrestrial globe for orientation.*"

Shipping Routes and Sea Cables

The Wirtschafts-Politische-Globus was a popular model in early-twentieth-century Germany, dotted with icons and names. A large sphere could display more than forty thousand items, demonstrating the flux of goods and people in the global economy. The globe was covered in a gracious web of shipping routes, the bloodstream of colonies, slavery, and trade. Busy economic centers are located at the origin of these lines, with distances between them marked in miles or sailing days. Another web is formed by undersea telegraph cables. The first one was laid in the 1850s; a few years later all continents were interconnected. For a hundred years, it was the most important long-distance communication network. In the 1930s, with the advent of radio communication, globe makers added antenna icons on their globes to mark the position of radio stations.

Zeppelins and Caravans

The Berlin-based globe maker Paul Oestergaard created the Welt-Verkehrs-Globus from the 1910s onwards. The monumental floor globe, with its forty-eight-centimeter sphere was the eye catcher on ocean steamers of Norddeutscher Lloyd, keeping passengers occupied on their five-day Atlantic journey. The globe displayed the latest transportation connections, such as railroads and shipping lines. In the 1930s, when the German airships Graf Zeppelin and Hindenburg began transatlantic service, their flight routes soon appeared on globes. In the 1940s, intercontinental aviation routes were added. The automobile highway system, constructed in the 1950s, is rarely drawn on globes, as it would take up too much space. Up to the 1970s, one could still find dotted lines in the Sahara Desert on certain globes, marking the traditional camel caravan trade routes.

Knowledge is Power!
Newspaper advertisement / Globe maker Paul Oestergaard / *Berlin, 1910s*

70
Columbus Verlag in Stuttgart created this small fourteen-centimeter globe in the 1960s. Although it is a small model, it still demonstrates the thoroughness of the German globe maker. It presents a very detailed cartography by W. Kaden, in the tradition of the earlier world traffic globes. The map features many shipping lines, drawn as blue bundles between port cities. The thickness of each bundle indicates the importance of the traffic between the ports. The sphere is mounted on a hardwood stand in the shape of a ship.

AIR-AGE

Airways to Peace

Bicycle mechanics Orville (1871-1948) and Wilbur Wright (1867-1912) started the century of aviation in 1903, when their self-crafted machine took off from Kitty Hawk Beach. Aviation proved its potential during the First World War. Ten years later, American Airlines had already transported over one million passengers. The growth of civil aviation coincided with global warfare in the 1940s, highlighting the interconnectedness of distant places. Cartographers responded with Air-Age geography: instead of viewing oceans as dividers between continents, they placed the North Pole as a unifying element at the center of their maps, focusing on connections and proximities. In 1943, the MoMA in New York organized an exhibition titled *Airways to Peace*, confronting the public with a globalizing world by showcasing these new types of maps and globes.

Flight Routes

To prepare American children for the new era, Air-Age education was introduced into the curriculum in the 1940s. Cram responded to this initiative with Air-Age globes, using the slogan: *"Today's School Children Will Live in Tomorrow's Air-Age World"*. Globes were promoted as the best tools to visualize relationships between places on Earth. The fact that the shortest flight route sometimes goes over the North Pole is incomprehensible on a traditional flat map. Air-Age globes helped air travelers plan their trips: *"A Cram Globe shows the hundreds-of-thousands of miles of air routes that will make all nations neighbors, tomorrow"*. Air-Age globes often rest on a cradle mount or clear view mount, allowing one to hold the Earth and observe it as if flying in an airplane. On these globes, the new flight routes spun a dense web of lines, similar to the shipping routes that preceded them.

Distance Finder

In the same period, Rand McNally released its own Air-Age Globe, with the intention of increasing public awareness of the shrinking world. The sphere rests on a cradle mount, a black ring made of glass, as a solution for the metal scarcity during the war. A transparent Air-Distance Measuring Tape to measure flight distances between airports came with the globe. Replogle offered an Air-Ways-Globe, aimed at the surging number of air travelers, with possible flight routes indicated in red lines. A complementary cardboard arc, the Distance-Finder, made it possible to measure flight time and distance. The globe also features a cradle mount, made from pressed cardboard, another solution to save on scarce materials in the postwar period. The air traveler could pick up the sphere from its cradle to examine the flight routes unobstructed [no. 173].

You and Postwar Air Transportation
American Airlines
Promotional
Poster / U.S., 1940s

71

Rand McNally from Chicago created this fascinating Air Globe for American Airlines in 1944, with paper map gores and a black glass ring to save metal during the war. The twelve-inch globe features blank map gores, showing only the names of airports, as if one were flying above the clouds. The cluster of airport names forms the outline of the landmasses. The globe represents a new world, interconnected by air travel and liberated from geographic obstructions. The Air Globe was used for an American Airlines advertising campaign, with the slogan: *"All Mother Earth's Children Live on the Same Street"*.

ADVERTISEMENT

GLOBES

Desired Behavior

Austrian-American Edward Bernays (1891-1995), nicknamed 'the father of public relations', was a nephew of Sigmund Freud (1856-1939). He turned his uncle's theories into a model for marketing and propaganda, using crowd psychology to direct the masses towards desired behaviors. His groundbreaking campaign in the 1920s promoted female smoking, a taboo at the time, by branding cigarettes as 'torches of freedom' to appeal to the first feminist wave. Newspaper advertising was followed by the first radio spot in the 1920s, the first television commercial in the 1940s, and the first internet banner in the 1990s. In the twentieth century, entire buildings, vehicles, and clothing were covered in commercial messages. Globes did not escape this trend: most globe makers offered customized globes for advertising, adorned with brand logos.

Brands and Logos

The Buffalo Evening News ordered a custom-made globe from Cram in the early 1930s. A standard seven-inch model was transformed into an advertisement globe by simply glueing a round piece of printed paper to the metal base, carrying a straight-forward message: "Buffalo Evening News, Western New York's Greatest Newspaper". The newspaper offered the globe for thirty-nine cents as an incentive for new subscribers [no. 172]. In the same period, New Jersey tin toymaker J. Chein made advertisement globes. They customized a bank globe for the Chevrolet company by lithographing the logo and a slogan on its metal base: "Chevrolet - The Symbol of Saving for 27 years". Pan Am and Air France had their logos printed on inflatable globes. These cheerful objects aimed to seduce travelers into buying airplane tickets by fueling their desires for exotic places globe [nos. 16-17].

Shampoo and Toothpaste

In the 1950s, globes were used to incite the public to buy products. A Palmolive-Colgate campaign offered a globe at reduced price when purchasing shampoo or toothpaste [see image above]. The advertisement read: "Worth $4.95 - Colgate-Palmolive makes it possible for you to get this 12-inch World Globe for only $2 plus carton from Colgate Dental Cream or Halo Shampoo!" Oil companies used inflatable globes to attract customers to their gas stations: "Approved by educators, this handsome eight-color, library-size Hammond Globe is heavy, durable vinyl ... inflates to full fourteen inches ... spins in sturdy metal holder ... available exclusively through Amoco Dealers ... only $3.50." During the moonshot era, lunar globes were used as incentives to buy television sets: "Free Moon Globe with the purchase of any Zenith Color TV - Enjoy all the thrills of future Apollo moonshots on a Zenith Color TV!"

A World of Your Own!
A globe as a premium for toothpaste / Palmolive-Colgate Advertisement / U.S., 1958

72
Weber Costello in Chicago produced customized advertisement globes alongside their educational globe line. This 1950s model was made for Goodyear, the largest American rubber tire factory. it was designed to attract attention at the reception desks of car showrooms. The Goodyear Globe is based on Weber Costello's standard ten-inch Peerless Globe, in this case, a black ocean version. The object promotes Goodyear's products in a clever way by matching the black map gores with a surprising base: a miniature black rubber tire, creating an eye-catching combination.

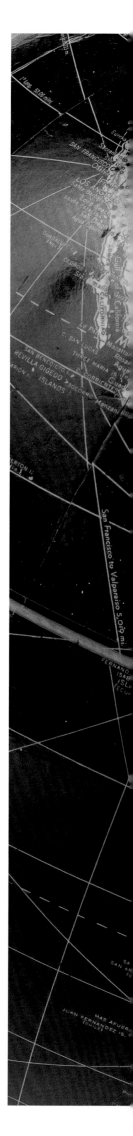

"Knowledge is Power! Knowledge of our Earth is of the utmost importance, especially in our busy time of steam and electricity, which spans all distances. Today, when the telegraph brings news from the furthest corners of the Earth in a very short time, it is essential to have a really good terrestrial globe for orientation."

Paul Oestergaard, 1910s
Owner of Columbus Verlag
Berlin globe maker

Delta Air Lines
Stewardess
showing a flight
route on a globe
in a mock-up
interior of a
new Convair 880
jet airliner /
Promotional
picture /
U.S., 1958

BANK
GLOBES

Money in the Bank

Charles Dow (1851-1902) co-founded the *Wall Street Journal* and initiated the Dow Jones Stock Index in 1898. The index rose from fifty to ten thousand points in the twentieth century. However, the journey was not always smooth, as seen in the 1929 Wall Street Crash and during the subsequent Great Depression. As a result, stabilizing mechanisms were created at the Bretton Woods Conference in 1944, such as the International Monetary Fund and the World Bank. However, these mechanisms did not prevent new significant crashes, like Black Monday in 1987 or the Financial Crisis of 2008. The volatility of stocks often scares people out of the market, making them prefer to keep their money in safer places, such as a clay container, which throughout history morphed into a pig, a symbol of fortune. In the twentieth century, some coin containers even evolved into a globe, a symbol of power.

First Pennies Then Dollars

The oldest American bank globes are miniature models, dating from around 1900. They consist of two cast-iron hemispheres with a very limited embossed map. A coin slot was cut out, and the two parts could be unscrewed to release the coins. Due to their small size, few coins would fit [no. 82]. This did not stop Enterprise Mould Manufacturing Co., a Philadelphia kitchenware factory, from decorating a cast-iron bank globe with the encouraging text: *"First Pennies Then Dollars"*. In the same period, German toymakers turned their standard miniature tin globes into bank globes by cutting a slot in the hollow sphere. Sometimes a hollow base was added to enlarge the coin compartment. Their cartography was more detailed, thanks to the color-lithographed map, making it an excellent object to teach children geography and finance simultaneously [no. 83].

As You Save so You Prosper

Most tin toy manufacturers produced popular metal bank globes. In the U.S., there were Ohio Art and J. Chein, in Europe Michael Seidel and Chad Valley. The three to five-inch spheres featured color-lithographed cartography with a coin slit. At the bottom, the disc-shaped base had a key-operated lid. It was often customized with a decoration or an energizing slogan, such as: *"As you save so you prosper"* [no. 86]. Even serious globe makers produced bank globes in the 1930s. Denoyer-Geppert offered a two-and-a-half-inch bank globe, with paper map gores and a coin slit in the Pacific. It was customized by adding a red-enameled cap to the North Pole, displaying a commercial bank's logo [no. 81]. Columbus Verlag in Berlin advertised bank globes with an educational message: *"A Meaningful Gift for the Child: It Plays, It Saves, It Learns"* [see picture above].

A Meaningful Gift for Children
3 Things: The Child
It Plays, It Saves,
It Learns /
Columbus Verlag /
Berlin, 1930s

73-74-75
MK-Tuote was founded in Helsinki in 1959 by Heikki Tavela (1932-2013) and specialized in plastic household objects. This striking Maapallo Coin Bank (Finnish for globe) from the 1960s is cast from translucent plastic in an electric-blue, red, or green color, with the continents spray-painted in gold. The ten-centimeter sphere is customized by adding the name of the bank on an exchangeable plastic strip attached at the equator. Many versions exist in flashy colors, each carrying the name of a different financial company from all over the world. MK-Tuote was so successful that it received the Export Award of the Finnish president in 1970.

MISSION

God on Earth

James Cash Penney (1875-1971) opened his first store, called The Golden Rule, in 1902, and his thousandth store sixteen years later. His chain of department stores became one of the largest in the U.S. in the twentieth century. Although the son of a Baptist preacher from Missouri, Penney only came to faith at a later age. After converting to Christianity and becoming a Born-Again Christian, he established the Golden Rule foundation to promote Christian values. In the late 1930s, the Foundation ordered a Mission Globe [no. 76]. Usually, these were standard tin bank globes made by one of the large tin toy factories, customized with a Christian slogan on the base or even on top of the cartography. The globe symbolized the spreading of God's Word across the Earth. And, strategically placed in a shop or public building, it was a promotor of faith and a collector of money.

Protestant Missions

The Anabaptist Mennonites are followers of the Dutch preacher Menno Simons (1496-1561). They immigrated to America in the eighteenth century. In the 1960s, the Mennonite Board of Missions in Elkhart, Indiana, ordered a custom-made tin bank globe from the Ohio Art Company with its name printed on the globe's base, to collect money for its activities [no. 85]. In the same period, the Rheinische Missionsgesellschaft, a German Evangelical missionary founded in the nineteenth century, ordered a custom-made bank globe from the Finnish company MK-Tuote. Its model featured an electric-blue translucent plastic globe with golden cartography, the same model used to promote commercial banks. In this case, the Word of Jesus adorns the white equatorial ring: *"Ihr werdet meine Zeugen sein - Gaben für die Rheinische Mission".*

Catholic Missions

The Society for The Propagation of the Faith is a Catholic charity founded in France, in the early nineteenth century. It sends missionaries across the world to spread God's Word. In the 1950s, it ordered a mission globe from the American Map Company, to collect money for its cause. The custom-made map of the tin lithographed globe reveals the society's rather naive worldview. Each continent has a special color, serving as a metaphor for its inhabitants, with explanations provided in the surrounding oceans: *"... Green is for Africa. It is the sacred color of the Muslims for whom we pray, and for all those who live in its green forests ... Red is for America because the Red Men first founded the continent ... Yellow is for Asia, It stands for the morning light of the East and the cradle of world civilization ... White is for Europe, because the White Shepherd, our Holy Father, lives in Rome..."* [no. 84].

Spread the Word
Geographical Training of Catholic missionaries, searching for destinations on a globe / *Rome, 1930s*

76
The J.C. Penney's Golden Rule Foundation distributed a mission globe around 1938. The map of the four-inch tin globe features biblical texts, such as: *"Do unto others as you would have them do unto you",* and slogans of the organization, such as: *"The world is now one neighborhood - The Golden Rule would make it one brotherhood".* The base includes poems by popular poet Grace Noll Crowell (1877-1969), who was proclaimed 'American Mother of the Year' by the Golden Rule Foundation: *"Eat your bread in solitude and you will be fed but share your loaf if you would know the flavor of good bread".*

RADIO

Short Waves

Italian engineer Guglielmo Marconi (1874-1937) succeeded in sending long-distance radio messages over the Atlantic Ocean around 1900. Radio waves can travel long distances because the ionosphere reflects the waves back to Earth. Until then, transatlantic communication was only possible by telegraph, using deep-sea cables, which were featured on the globes of the period. The sea cables had a strategic flaw: two-thirds of the network was in British hands, giving them an intelligence advantage over other countries that used the network. To circumvent British eavesdropping, the Dutch government built a giant shortwave radio station in the 1910s, consisting of six antennas, each 212 meters high. It enabled them to avoid using the British undersea cables and communicate directly by radio to Indonesia, a Dutch colony at the time.

Fireside Chats

By the 1930s, the Golden Age of Radio had begun. Half of all transcontinental communication was now wireless, and two-thirds of American households owned a radio. A radio was the central element of the living room, a heavy piece of wooden furniture where families gathered around. It was the main source of information and entertainment, before the arrival of television. U.S. President Franklin D. Roosevelt (1882-1945) created a communication revolution by using the new medium to connect directly to American voters in their living rooms. His radio talks, the so-called 'fireside chats', boosted globe sales. When explaining international politics, he told his listeners: *"... and now bring up your globe!"*. On the globes of that period, antenna icons and call letters were added to mark radio stations, so one could see where these faraway voices broadcasted from.

Tallinn to Tokyo

In the 1950s, bulky radio tubes were replaced by compact transistors. The transistor radio was the precursor to all portable media, such as the Walkman in the 1980s or the Ipod in the 2000s. Some radio manufacturers built a small radio into a globe. Japanese Metex created the Fleetwood Transistor Six in the early 1960s, a six-inch golden plastic globe, mounted on a black Bakelite base. The radio is integrated into the globe, with a loudspeaker at the South Pole, a volume disc at the North Pole and a tuner slide along the International Date Line in the Pacific. Its design expresses the globalizing effect of radio [no. 88]. The RAHU-87 by Tallinn Punane RET Radio Works in Soviet Estonia was a 1980s radio built into the base of an illuminated plastic globe. The radio is tuned by rotating the globe, a surprising product from a country where at the time foreign broadcasts were jammed [no. 89].

S.O.S
Children follow the rescue operation of their aunt, aviatrix Amelia Earhart, on the radio and on the globe / *U.S., 1937*

77
Radio-Célard, a French radio maker from Grenoble, created this fine example of a radio globe, called Captemonde, around 1955. The paper gores feature interesting pictorial cartography, with airplanes, clippers, sea monsters and mermaids drawn in the oceans. Its graphics are executed in a typical 1950s joyful French style, with fresh colors. The twenty-centimeter sphere is held by a metal double ring that looks like a full meridian but is actually a circular antenna to receive the radio signal. The antenna is mounted on a plastic base that houses the radio device and includes an integrated tuner knob and volume slide.

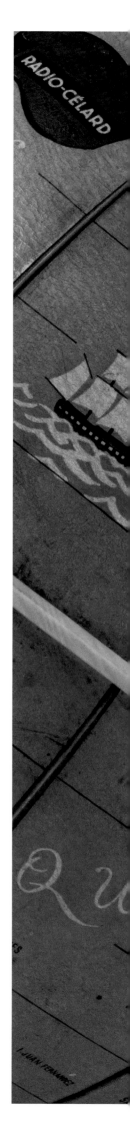

CLOCK

Train Tables

Phileas Fogg, the main character in Jules Verne's book *Around the World in Eighty Days*, makes a bet that he can travel around the globe in eighty days. One day too late, he arrives in London, only to realize at the last minute that he won the bet because he saved a day by traveling around the world counterclockwise. Standardized time did not yet exist in the nineteenth century. In each village, the church clock was set to twelve when the Sun was at its highest point. As a result, every village had the correct solar time, but each village had a different time. This became problematic when the railways were invented: trains need a synchronized timetable. In 1847, British Railways set the standard time for the entire U.K., known as the Greenwich Mean Time, which was confirmed by the International Meridian Conference forty years later as the basis for the international time system.

Time Zones

The conference divided the Earth into twenty-four official time zones and drew a date line in the Pacific Ocean. It took another forty years before most countries had adapted to the new system. In the interwar period, a number of clock globes were created to illustrate the connection between the rotation of the Earth and the time zones. Some models have a terrestrial sphere that rotates around its axis in twenty-four hours, while others have hour markers that rotate around a fixed globe. Globe maker Replogle integrated a Lux Rotary Mystery Tape Measure Clock in the circular base of a standard globe. This spring-operated clockwork, a patent by clockmaker Herman Lux (1896-1979), was advertised as *"The First Different Clock In Centuries"*. It has twenty-four-hour markers that rotate around the base, indicating local time in each part of the world [no. 92].

Geokrono

Italian Umberto Nistri (1895-1962) pioneered aerial cartography in the Italian Air Force during the First World War. After the war, he founded Ottico Meccanica Italiana in Rome, a company that made photometric instruments. In the 1930s, Nistri developed the OMI Alpha Cryptograph, a secret rotor-coding machine used by Italy to send messages during the Second World War. OMI also developed other novelty instruments, such as a clock globe called Geokrono. It is made of two metal hemispheres, covered with a colorful lithographed map image. The equator is split by a white ring, which has twenty-four time markers. The ring turns around the globe in a day, allowing the observer to compare the hour in every time zone. The sphere is mounted on a heavy metal stand, and it comes with a complementary key to manually wind up the mechanism of the clock globe [no. 91].

The News Hour
Douglas Edwards,
first CBS TV anchor,
in his studio with
a globe and time
zone clocks /
U.S., 1948

78
Dietrich Reimer, a well-known German globe maker from Berlin, made this clock globe in the 1930s. Its fifteen-centimeter sphere is covered with paper map gores that display detailed cartography. The sphere is connected to an electric motor hidden in the chromed metal base which rotates it in a full circle over twenty-four hours. The local time of any place on Earth can be read from the hour markers on the horizon ring. The model is a so-called capped globe where a metal shield covers half of the sphere, marking the dark night side of the Earth. Integrated into the base are displays for the months and the seconds.

OFFICE

Scientific Management

Stenographer William Henry Leffingwell (1876-1934) was the pioneer of Scientific Management, the theory of the optimization of office work. In the early twentieth century, he wrote influential textbooks covering a range of subjects, such as space planning, the use of typewriters, and the efficiency of daily routines. In the twentieth century, the economic activity shifted from agriculture and industry to services. As a result, large parts of the population spent the best part of their lives in an office environment. Traditionally, globes have been a common feature in an office. They may serve a practical purpose for finding geographical locations in an international business company, but more often, they serve a symbolic function, to demonstrate the company's sophistication. Office inventories often include various instruments of efficiency, some cleverly disguised as globes.

Pencils and Calendars

For an efficient administration, it is crucial to keep the date and follow the calendar, a mathematical system based on the movement of the Earth. Its spin defines the rhythm of day and night, its orbit around the Sun defines the length of the year, and the inclination of its axis defines the seasons. The relationship between the Earth's movement and time is illustrated by a calendar globe. A good example is a small bronze globe, which displays the date as a digit in a small cut-out window. Another office amenity is the pen tray globe. Sometimes, a wooden base is crafted into a pen holder, as seen in a model by Cram. These pen tray globes can also take on playful designs for children, such as the light-blue tin globe French toy manufacturer Jouets TMF produced in the 1950s, which floats on a gold-colored steel wire base, folded as an elegant pen holder [no. 125].

Zonemaster

American Inventor Fred G. Burg (1896-1995) was a Jewish emigrant from Bohemia who founded the Burgmaster Corp. in Gardena, California, a firm specializing in mechanical instruments. Burg obtained more than one hundred patents; among them is the Zonemaster, also known as the Presidential World Globe Time Zone Clock. This device features a six-inch brass relief globe with an integrated clock that displays the local time. The brass base holds a name roll, listing one hundred sixty-six world cities. By turning a knob on the base, the outer disc turns to a position that shows the time of the chosen city relative to local time. A red light indicates whether it is day or night in the selected location. Fred G. Burg promoted the Zonemaster as a useful device on the desk of someone who wants to be on top of the world: *"A Magnificent gift for the Jet-Age man!"* [no. 90].

The Zonemaster
Fred G.Burg
demonstrates his
Presidential World
Globe Time Zone
Clock / *U.S., 1965*

79-80
Columbus Verlag in Stuttgart published this fine pair of office bookend globes in the 1950s. Many globe makers used bookends to continue the longstanding tradition of selling globes as a pair: a terrestrial globe and a celestial globe. Bookend globes look great on the bookshelves of a study room. This particular model consists of eleven-centimeter miniature spheres with detailed cartography, a scaled-down version of Columbus Verlag's larger globes. The spheres are fixed to modernist-styled book supports made of curved plywood, featuring a dark surface coating that contrasts with the light exposed veneer lines of the edges.

More Bank globes

81

Denoyer-Geppert from Chicago produced small, customized bank globes in the 1920s. This two-and-a-half-inch miniature globe has detailed paper map gores and a coin slit in the Pacific Ocean. It is customized by adding a red-enameled cap bearing the bank's logo at the North Pole. The little sphere rests on a metal pin fixed to a wooden base.

82

The Grey Iron Casting Company of Mount Joy, Pennsylvania, U.S., made miniature bank globes around 1900. The three-inch sphere consists of two cast-iron hemispheres, which can be unscrewed to collect the coins. The continents are cast in relief. The stand is also made of cast iron. All parts are painted red, though other colors were available.

83

Around 1900, German toy makers turned their miniature tin globes into banks by simply cutting a slot in the sphere of an existing model. Since not many coins fit in this eight-centimeter sphere, a hollow aluminum base was added. These miniature models feature a detailed cartography, thanks to the sharp color-lithographed map images.

More Mission Globes

84

The Catholic Society for The Propagation of the Faith ordered a mission globe from the American Map Company in the 1950s, to collect money for its cause. Each continent of this five-and-a-half-inch tin lithographed globe is depicted in a bright color, serving as a metaphor for its inhabitants, as explained by the texts printed in the oceans.

85

The Mennonite Board of Missions and Charities in Elkhart, Indiana, ordered a custom-made bank globe from the Ohio Art Company in the 1960s to collect money for its mission activities. The four-inch lithographed tin terrestrial sphere has a coin slot and is fixed on a tin base, on which the name of the Mennonite Mission is printed.

86

 J. Chein was a versatile tin toy factory in Burlington, New Jersey. The company manufactured small globes, often custom-made for clients. In 1939, J. Chein produced this bank globe as a souvenir for the New York World Fair. The four-inch lithographed tin sphere had silver oceans, colored banners and the Fair's logo printed on its tin base.

More Radio Globes

87

Radio Manufacturer Nipco in Tokyo created this twelve-centimeter radio globe in the 1960s. The sphere has paper map gores with unique cartography that depicts various ocean depths. The same map image was used for other Japanese globes. The sphere is mounted on a metal meridian and a plastic base, which contains a built-in radio.

88

The Japanese Metex Radio Company made this Fleetwood Transistor Six, model NTR-6G, in the early 1960s. The six-inch gold-colored plastic sphere is mounted on a Bakelite base. The radio components are integrated into the globe, with a loudspeaker at the South Pole, a volume disc at the North Pole, and a tuner slide on the International Date Line.

89

In the 1980s, Tallinn Punane RET Radio Works in Soviet Estonia created this globe radio called RAHU-87. The radio, incorporated in the plastic base, can be tuned by turning the twelve-centimeter illuminated sphere. It has paper map gores in Russian. The globe radio was available in both terrestrial and celestial models.

More Clock Globe

90

Burgmaster Corp. in Gardena, California specialized in office machines, such as this 1960s Zonemaster, also known as the Presidential World Globe Time Zone Clock. It is a six-inch brass relief globe with a clock displaying local time. When choosing a world city from the name roll in the base, the outer disc indicates the local time in that city.

91

Ottico Meccanica Italiana in Rome designed the Geokrono in the 1940s. It is made of twelve-centimeter metal hemispheres, lithographed with a colorful map. At the equator, a white ring shows the hours of the day. It slowly turns around the globe in twenty-four hours, allowing the observer to see the hour in every time zone simultaneously.

92

Replogle, the Chicago globe maker, produced this clock globe in the 1930s. The ten-inch terrestrial sphere is fixed to a full aluminum meridian. It stands on a circular metal base with an integrated Tape Measure Clock. The twenty-four-hour markers on the circular dial rotate around the base to indicate the time in different parts of the world.

SPACE

Men on the Moon

Neil Armstrong (1930-2012) was the first man to set foot on the Moon on July 21, 1969. When he stepped out of the Apollo 11 Lunar Module, he uttered the historic words *"That's one small step for man, one giant leap for mankind"*. The promise of President John F. Kennedy to put a man on the Moon before the end of the decade was realized.

The moonshot turned into a phenomenon during the same period when television emerged as a mass medium. A large part of the world's population sat breathless in front of small black-and-white TV sets, gazing at the grainy images beamed from the Moon that showed Neil Armstrong descending the spaceship's ladder. Many television channels broadcasted a daily Moon program, with live reports covering the exiting journey of the astronauts. A giant lunar relief globe was the main feature in every TV studio over the world. The CBS company had bought the exclusive rights to a six-foot globe, custom-made by Rand McNally.

Walter Cronkite (1916-2009), the popular anchorman, covered the moon landing for CBS TV. He used the globe as a visual demonstration model to explain all aspects of the Apollo 11 voyage. In this picture, the six-foot globe was used as a prop for a photo shoot of the Apollo 11 crew. (from left to right: Buzz Aldrin (1930-), Michael Collins (1930-2021) and Neil Armstrong (1930-2012))

In the twentieth century, new theories about the origin of the universe emerged. Popular fascination was triggered by science fiction as well as by the Space Race, which culminated in the historic moment of the first moon landing, when the whole world watched the first man walk on the Moon on their television sets. As a result, Moon globes found their way to an enthusiastic audience. Traditional celestial globes also continued to be used for navigation, even in spaceships.

CELESTIAL
GLOBES

Galaxies and Candy Bars

Frank C. Mars (1882-1934) from Minnesota was the most success-ful candy maker of the twentieth century. He invented the first filled chocolate bar in 1923 and called it Milky Way. It was a ref-erence to the galaxy, the hazy band of stars that looks like spilled milk and can be seen at night. At the time, a fierce battle raged among astronomers about the Milky Way, the nature of the spiral nebula and the size of the universe, the so-called Great Debate. It was covered by newspapers and closely followed by the public. Edwin Hubble (1889-1953), the most prominent astronomer of the twentieth century, settled the issue in the 1920s when he proved that our galaxy was just one of many. As a consequence, the uni-verse had to be much larger than assumed. It was a great time for candy makers to sell Milky Ways and for globe makers to sell celestial globes.

Celestial Misconceptions

Historically, globes came as a pair: a terrestrial and a celestial one. Celestial globes depict the stars as observed in the night sky. They existed already in antiquity, long before terrestrial globes. It is easy to understand why: when you look up at night, half of the universe is directly visible, but during the day, ninety-nine percent of the Earth is hidden behind the horizon. Scientifically, a celes-tial globe is a misconception. First, the shape of the universe is not a sphere with stars on its surface, all positioned at an equal distance from Earth. Second, we perceive the universe as though we are inside a hollow dome, however, a celestial globe presents the universe as if viewed from the outside, resulting in the star map being depicted mirrorwise. In the 1940s, Robert H. Farquhar solved this issue by designing a transparent celestial globe, which makes it possible to look inside out [no. 96].

Star Magnitudes

Globe makers often relied on external astronomers for their ce-lestial maps. For instance, C.S. Hammond & Co., for its 1940s ce-lestial globe, commissioned Commander S.E. Stubbs of the Royal Navy to draw a star map [no. 104]. Celestial globes come in many styles. They display astronomical knowledge through a variety of icons, such as star magnitudes on a scale of one to five. The uni-verse can be depicted in black, blue, greenish or white, with stars represented as gilded or black dots, the Milky Way as a white swirl-ing cloud [nos. 5, 30, 159]. Star maps are either printed on paper or lithographed on metal, such as the small 1970s set called Worlds Unlimited by Scan-Globe, featuring Earth, Moon and Mars globes [no. 103]. For the celestial globes, the same stands were used as for the terrestrial ones: a Bakelite base, a wooden cradle base, a plastic mono leg, a steel wire tripod, and so on.

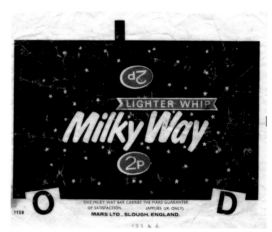

Milky Way
The candy bar named after an astronomical phenomenon features a celestial map on its wrapper / Mars Ltd., Slough, England, 1970s

93
Columbus Verlag in Berlin and Stuttgart published a deep-blue Himmelglobus in the 1920s, compiled by astronomer Dr. Johannes Riem (1868-1943) and cartographer C. Luther. Stars are represented as yellow icons, with magnitudes from one to five, interconnected by black lines to form constellations. The galaxy is depicted as a white cloud. The globe was sold on various mounts; this 1950s model has a thirty-four-centimeter illuminated glass sphere, embraced by a black-aluminum meridian that transitions into an elegant, streamlined base. The same model was available as a terrestrial globe [NO. 152].

CONSTELLATION
GLOBES

Physicists and Superheroes

German-born Albert Einstein (1879-1955) was one of many phys-
icists in the early twentieth century who promoted groundbreaking new concepts about the universe. His 1915 *Theory of Relativity*
introduced concepts such as time travel and parallel dimensions.
A few years later, Belgian Jesuit and physicist Georges Lemaire
(1894-1966) was able to reconcile religion and science when he
postulated the Big Bang Theory. This theory suggests that the
universe originated from a singular point, more than thirteen billion years ago, and has been expanding ever since. These exciting
new scientific theories inspired popular culture. Comic book artist
started a new genre in the 1930s with superheroes from faraway
galaxies, such as Superman, who was born on planet Krypton and
possesses supernatural powers, including the ability to travel in
time.

Zodiac and Horoscope

Superheroes stand in a long tradition of fascination with the universe. Already in ancient times, people tried to grasp the mys-
teries of the night sky by joining stars into the constellations and
drawing them as mythical figures and animal shapes on their celestial globes. These constellations of the zodiac were believed to
have cosmic powers that influence the fate of humans on Earth,
an idea continued in the pseudoscience of astrology and its horo-
scopes. In the 1920s, the International Astronomical Union used
the constellations as a basis for an official mapping of the universe, listing eighty-eight constellations as rectangular territories
in the sky. Some globe makers adopted this new system, while
others retained the mythological figures floating on a background
of golden stars to create romantic constellation globes [nos. 30, 94,
159, 166, 168].

Mythical Animals

Vallardi, the Italian publishing house, created a remarkable constellation globe. Its map gores were designed by Italian astronomer Ernesto Sergent-Marceau (1832-1897). The pale-green sphere,
adorned with classic illustrations of the zodiac, balances on top of
a modernist brown Bakelite fin-shaped stand. The Danish Scan-
Globe created a twelve-inch Constellation Globe in the 1970s,
with a star map by Karl F. Harig (1940-). The acrylate sphere displays a blue celestial globe with yellow stars. However, once the
internal light is switched on, a colorful drawing of the constellations comes to life, with green and red animals and mythological
figures floating in a purple universe. The globe came with an exuberant curved transparent Plexiglas stand, that was also used for
terrestrial globes by Scan-Globe and Replogle during that period
[no. 105].

Superman
Cover of the
comics book
series *No. 6*
by Jerry Siegel
and Joe Shuster /
U.S., 1940

94
Rand McNally from Chicago created this
twelve-inch constellation globe in the
1950s. Its paper gores feature a map drawn
by astronomer Oliver J. Lee (1881-1964). The
stars are bright yellow, with magnitudes from
one to five. Classical constellation figures
are drawn as light-blue lines on a dark-blue
background, in a fluid contemporary style.
The sphere is fixed to a gimbal metal stand,
allowing it to be rotated and observed from
all directions. The company sold a smaller
nine-inch globe with the same constellation
map, mounted on a Bakelite stand, called
The Galileo [NO. 168].

PLANETARIUM
GLOBES

Beaming the Universe

Armand Neustadter Spitz (1904-1971), an American journalist and lecturer, used his communication skills to promote popular astronomy in the U.S., before becoming the education director at the Franklin Institute in Philadelphia. A perfect tool to build enthusiasm for astronomy is a motorized planetarium that projects the night sky onto a dome ceiling of a public auditorium. As this is an expensive device, Spitz developed a portable planetarium in the 1940s, selling for only $500. The story goes that Albert Einstein suggested Spitz to use a dodecahedron shape. It approximates a sphere, but its flat surfaces facilitate punching pinholes more easily. An interior lamp beams light through the pinholes, projecting the universe on the ceiling. Spitz started his company in 1947, which is the world's leading planetarium manufacturer still today [see picture above].

Spitz Planetarium
Armand N. Spitz
demonstrates
his A1 planetarium
to children /
U.S., 1950s

Spitz Sky Zoo

Spitz's next step was an affordable home planetarium. In 1954, he created the Spitz Junior Planetarium, produced by Harmonic Reed, a manufacturer of toy musical instruments. It was a big success, with Spitz selling one million copies at only $14.95 each. The device is a small plastic globe, adjustable on a Bakelite base to adapt the projection angle. Light is beamed through three hundred pinholes on the globe's surface, projecting the stars onto the ceiling of a living room. An update was the Spitz Sky Zoo, which could project images of constellations. Spitz later commented: *"I never expected to make any substantial contribution to astronomy or science, but what greater satisfaction can I have than to have one very famous astronomer tell me that he gained his first interest in astronomy through viewing a Spitz planetarium when he was a small boy"* [no. 107].

Farquhar Space Laboratory

Robert H. Farquhar was another hobby astronomer from Philadelphia with a similar ambition to educate the general public about the universe. In the 1940s, he designed a transparent celestial globe [no. 96]. In the 1950s, the Farquhar Transparent Globe Company developed the Farquhar Space Laboratory, a demonstration device for schools. The planetarium consists of a Plexiglas half dome, available in nine, fourteen or twenty-foot diameters. Students sit under the dome while light is beamed through a set of eight transparent globes, each with different map drawings. This setup projects stars and constellations of the universe onto the dome and can also be used to project all kinds of maps of the Earth. The $5,000 mobile planetarium could be taken apart and transported to other schools to share costs.

95
Robert H. Farquhar in Philadelphia developed the Bowl of Night in the 1970s. It is a twelve-inch Plexiglas half sphere, a cross between a planisphere and a planetarium. When the instrument is set to a specific time and location on Earth, it reveals the nightly constellations visible at that particular place. The Bowl of Night was made in collaboration with cartographer Georg Lovi (1939-1993). Farquhar used the stick-figure constellations of A.H. Rey (1898-1977), a children's book writer and amateur astronomer, who wrote the popular book: *The Stars - A New Way to See Them,* first published in 1952 and still in print today.

NAVIGATION

Navisphère

Henri Julien Aved de Magnac (1836-1892), a captain of the French navy, invented the Navisphère, a celestial navigation globe. Since ancient times, sailors used celestial bodies to navigate, so a celestial globe equipped with measuring tools is a useful instrument for calculating positions aboard a ship. De Magnac proudly promoted his invention: *"... a nautical instrument of extremely simple construction and easily handled, by means of which nearly all the complex nautical problems may be solved in a few minutes, and without calculation, or, at least, with very little calculation."* The first model was produced in the late nineteenth century. During the twentieth century, various navigation globes were produced in different European countries. They were used until the end of the century, when satellite navigation systems made the Navisphère obsolete.

Tokyo-Berlin Non-Stop

The British Navy used a navigation globe, called the Starfinder Globe. It was made by Cary & Co and also sold in the 1920s as the Husun Star Globe by H. Hughes & Son. The instrument consists of a small celestial globe with yellow map gores for contrast at night. The globe is suspended in a vertical metal ring with latitude markings, inserted into a horizontal ring calibrated with compass degrees, and housed in a wooden box for easy transport around the ship. A Japanese version of the Navisphère was aboard the Tachikawa Ki-77 airplane, built during the Second World War to perform non-stop service between Tokyo and Berlin. Unfortunately, it did not prevent the prototype airplane from disappearing above the Indian Ocean during its maiden flight. A comparable navigation globe in a metal case was present on all ships and submarines of the Soviet Navy, until the 1990s [no. 114].

Demonstrations

Alexander Galt (1855-1938), curator of the Royal Scottish Museum in Edinburgh, designed a large celestial demonstration globe in the 1910s. It consists of a 150-centimeter glass sphere with the stars etched onto it. Looking through the transparent sphere, one can observe the stars as if standing on Earth. Inside the sphere is a scale model of the solar system that can be adjusted to show the movements of the celestial bodies. It was produced by the Michael Sendtner Fabrik für Präzisionsinstrumente in Munich, which built other copies for the Deutsches Museum in Munich, the Cranbrook Institute in Michigan, and the Franklin Institute in Philadelphia, hometown to globe maker Robert H. Farquhar. Given the similarities, the Sendtner Globe may have been inspirational to him when designing his own transparent navigation globe in the 1940s [see picture above].

Sendtner Globe
Transparent celestial sphere at the Cranbrook Institute of Science / *Michigan, 1937*

96
The Farquhar Transparent Globe Company in Philadelphia designed this Celestial Navigation Sphere Device 1X3 in the mid-1940s, in close collaboration with the U.S. Special Devices Center of the Office of Naval Research. The fourteen-inch Plexiglas sphere has the stars printed on it and rotates freely in a metal gyro mount, fixed to a wooden base. A terrestrial globe and a moon ball inside the sphere can be adjusted, and the position of stars during the day and the year can be observed. It was an immediate success; the U.S. Navy and Air Force ordered a few hundred for navigation training.

ASTRO
GLOBES

Afro-American Computers

Katherine Johnson (1919-2020) was an African-American mathematician at NASA. Following the social order in the 1950s, before the arrival of computers, men undertook the exciting design and engineering work, while women had to manually perform the complex calculations, often regarded as a tedious job. Considering the position of Black women in a period of racial segregation however, Johnson and her African-American female team had been entrusted with a critical task: they calculated the trajectories for American spacecraft such as the Mercury and Apollo missions that sent the first Americans in space in 1961 and the first men to the Moon in 1969. In the opposite picture, Johnson is making her orbit calculations with a Lafayette Astro Globe at hand, a scientific instrument to visualize the universe.

Human Computer
Katherine Johnson
with a Lafayette
Astro Globe on her
desk at NASA /
U.S., 1962

Lafayette's Astro Globe

The Lafayette Radio Electronics Company was one of the largest mail order electronics companies, selling television sets, hi-fi installations, and guitar effect pedals, such as the one bought by Jimi Hendrix (1942-1970) to create his famous psychedelic sound. Featuring on the cover of Lafayette's 1960 catalog is the Astro Globe, a striking object, designed by Hirose Hideo (1909-1981) of the Tokyo Astronomical Observatory [no. 97]. A smaller six-inch version was also available, the Torica Astro Globe, which functions in a similar way. The transparent acrylic celestial globe has the stars molded into its surface. In the center, a small Earth is attached to an axis. All parts are movable, allowing one to study the astrological positions throughout the year. The funky Torica, completely made out of plastic, is a crossover between a scientific instrument and a children's toy, making it the perfect Christmas gift [no. 106].

Farquhar's Earth in Space

The Farquhar Transparent Globe Company made an improved version of its Celestial Navigation Sphere Device 1X3 during the same period. The radical model was called the Earth in Space™ Celestial Globe. The twelve-inch globe was entirely made of transparent Plexiglas, including its double-curved base. Stars and constellations are serigraphed on the inner surface of the transparent sphere. A small Earth globe is mounted inside. Users can adjust the various parts to observe seasonal positions of celestial bodies. Just like Lafayette's Astro Globe, the fascinating Earth in Space™ Celestial Globe was promoted as a training instrument for pilots, sailors and astronauts. It is an iconic example of twentieth-century design, used by NASA for promotional photos to bolster the enthusiasm of the American public for space exploration [no. 164, p. 134].

97
The Lafayette Radio Electronics Company promoted this Astro Globe as a scientific instrument, even useful for professionals in space discovery: *"A Basic Aid for Students and Teachers of Astronomy - Navigators - Aeronautics - Astronautics"*. Through the transparent plastic sphere, with gold-painted stars molded into its surface and indications of the constellations, one can observe the stars as if standing on Earth. Inside, a miniature terrestrial globe, a red Moon, and a yellow Sun can be adjusted to study the positions and movements of the celestial bodies. The sphere is fixed to a stainless steel tripod stand.

"The Americans are making Moon globes, but they admit ours is the Rolls-Royce of Moons!"

Arthur J. Wightman, 1969
Owner of Lunasphere Productions
Globe maker in Penzance, U.K.

Astronaut John
Glenn looking
into a three-
feet Farquhar
Transparent
Celestial
Globe during a
training session
at the NASA
Aeromedical
Laboratory /
*Cape Canaveral,
Florida, 1961*

SATELLITE
GLOBES

Sputnik and Laika

Ukrainian engineer Sergej Pavlovich Korolev (1906-1966) developed rocket engines in the Soviet Union during the 1930s. After being falsely indicted and spending several years in Stalin's prison camps, he was brought back to Moscow during the Second World War to help develop war missiles. After the war, he directed a team of captured German scientists who brought their expertise on the V-2 rockets. Korolev became the father of the Soviet Space Program. On October 4, 1957, he made history when the Sputnik was launched, becoming the first satellite to orbit the Earth. It was a simple aluminum ball, fifty-eight centimeters in diameter, with four protruding antennas and a radio transmitter inside, sending beeps to Earth. Following his breakthrough, Korolev achieved another success with Sputnik 2, which carried a dog named Laika aboard.

Space Race

On April 12, 1961, Korolev's team launched the Vostok 1 spacecraft. It was piloted by Yuri Gagarin (1934-1968), the first man to orbit the Earth and return safely. To commemorate Gagarin's successful flight, the Soviets produced many copies of a small satellite globe. A little red rocket, attached to a steel wire, orbits a gold-colored plastic globe atop a cream-colored plastic base bearing the historical date [no. 112]. Two years later, Valentina Tereshkova (1937-) completed the Soviet achievement as the first woman in space. The Cold War rivalry between superpowers, the Soviet Union and the United States, ignited the Space Race. The Americans had to catch up: their first satellite, Explorer 1, was launched four months after Sputnik, and their first astronaut, John Glenn (1921-2016), followed the first cosmonaut Yuri Gagarin by eleven months.

Rings and Balls

American globe makers created various satellite globes in 1958 in an attempt to capitalize on this new phenomenon of satellites flying around the Earth. Replogle published a gyro-mounted terrestrial globe with a metal ball sliding on a metal ring to demonstrate the principle of a satellite orbiting around the Earth. Rand McNally also entered the Space Race by introducing its own satellite globe. In this case, a standard terrestrial globe with a metal horizon ring that can be hinged to demonstrate the angle of a satellite orbiting the Earth. The tin toy industry also produced several satellite globes, with objects orbiting around a metal globe. Some moved mechanically on a spring, while others were designed as agility games, such as the Satelit Rotaryo, made by the German Gescha factory, which uses hand movements to keep a satellite in orbit.

Sputnik
A Soviet engineer working on the first satellite to orbit the Earth / U.S.S.R., 1957

98
Denoyer-Geppert's first contribution to the Space Race was this Globe with Satellite-Orbit Demonstrator. Keith Olson, designer of many Denoyer-Geppert products, filed the patent in 1959. It was branded the Vanguard Globe, after the second American satellite, launched the year before. The twelve-inch terrestrial globe illustrates the orbit of a satellite by means of a hinged wooden horizon ring, fixed to a half meridian on the southern hemisphere. The wooden ring can be positioned at any angle to simulate the orbit of a satellite and show its various positions above the Earth.

SPACE NAVIGATION

GLOBES

Vostok and Mercury

Soviet cosmonaut Yuri Gagarin (1934-1968) was the first man to orbit the Earth on April 12, 1961. His Vostok 1 spacecraft was not more than a metal ball of only eight feet in diameter where he just fitted in. Gagarin's small stature was one of the reasons he was selected for the journey. The central feature of the control panel of the Vostok, right in front of the cosmonaut, was a navigation globe. The miniature globe adjusted itself to show the position above the Earth at any time. Until the end of the twentieth century, similar navigation globes were integrated into the control panels of all Soviet and American spacecraft. It is remarkable that a traditional globe was still used as a navigation tool in advanced spacecraft, but as early computer navigation systems were still susceptible to failure, a physical navigation instrument was considered vital.

Globus IMP

Globus IMP, the Russian abbreviation for Indicator of Position in Flight, was an important space navigation instrument built in the control panels of all Soviet spacecraft from 1961 to 2002, such as the Vostok, the Voskhod, and the Soyuz. The heart of the navigation instrument is a five-inch terrestrial globe, with blue oceans and orange landmasses. It moves freely within a transparent plastic dome, powered by a complex mechanical clockwork. A crosshair indicates the Nadir, the point on the surface of the Earth directly below the spacecraft. Above the globe is a rotating disc that indicates the longitude and latitude of the spacecraft's position in digits. The cosmonauts could instantly read their position in relation to the Earth on the navigation globe, especially at nighttime or if the viewport was not directed towards the Earth [no. 113].

EPI Globe

The EPI or Earth Path Indicator was the American version of a similar space navigator. It was used on board the Mercury spacecraft from 1962 onwards and produced by the Honeywell Corporation. Centrally positioned in the spacecraft's control panel, it consisted of a miniature terrestrial globe manufactured by the Farquhar Transparent Globe Company. The globe rotates behind a glass plate marked with a black arrow. It included dials for adjustment and a spring, as the astronauts had to manually wind up the clockwork mechanism. The revolving globe kept running for two days. As a precursor to the GPS system, the small globe indicated the location of the spacecraft above the Earth. This allowed the astronauts to verify their position in space or, upon their return, establish the point of descent and re-entry into the atmosphere [see picture above].

Earth Path Indicator
A spinning globe centrally built into the control panel of John Glenn's Mercury spacecraft / *NASA, 1962*

99

Salyut was the first space station that remained in permanent orbit around the Earth. Used by the Soviets between 1971 and 1986, it allowed cosmonauts to commute with a Soyuz spacecraft from Earth, dock at the space station, and stay for longer periods to perform experiments. On board was a celestial navigation globe called the Orbitometer. The eleven-and-a-half-centimeter motorized metal globe rotates within a plastic armature. Stars and constellations are etched onto the black surface, illuminated from the inside. The cosmonauts navigated by matching it to the stars they observed outside.

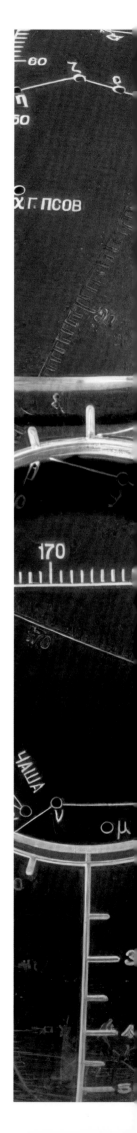

ROMANTIC MOON

GLOBES

Moonstruck

French film pioneer Georges Méliès (1861-1938) created the first science-fiction movie in 1902, *Le Voyage dans la Lune*, an adaptation of the novel *The First Men in the Moon* by the 'father of science-fiction literature', British author H.G. Wells (1866-1976). Méliès tells the story of people flying to the Moon, sixty-seven years before the real moon landing. Méliès was famous for his dramatic scenery and remarkable special effects. The movie featured the first film animation: a rocket landing on the face of the Moon. At the dawn of the twentieth century, Selenography, or the study of the Moon, became a hot topic. At that time, the only way to study the Moon was through a telescope; the moonshot of the 1960s, which would provide extensive scientific data was still far away. As a result, many romantic amateurs produced all kinds of speculative homemade lunar maps and globes.

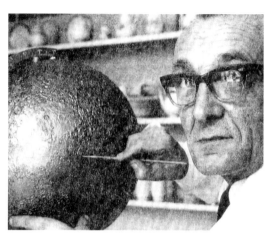

Alfred Schlegel
Pictured in his studio while sculpting craters on a Moon relief globe / *Germany, 1960s*

Edmund's Moon

Before 1492, terrestrial globes depicted only half of the world, as no European had yet seen or mapped America. Similarly, Moon globes were incomplete until 1959, because no human had observed the far side of the Moon yet. American Norman W. Edmund (1916-2012) sold military surplus after the Second World War, such as tank periscopes. He founded the Edmund Scientific Co. in New Jersey and produced optical instruments such as lenses and telescopes. Its 1962 trade catalog, beloved by science hobbyists, offered a twelve-inch relief Moon globe, *"molded of washable plastic, in three colors"*, that featured ocher craters and black plains, the lunar seas, as well as an empty far side, waiting for discovery: *"Reverse is blank and you can paint with watercolors to indicate present Russian satellite information, and update as more information is acquired."*

Schlegel's Moon

West German porcelain etcher Alfred Schlegel (1921-1997) became fascinated by space exploration while working on the V2-rocket program. After the war, Schlegel became a hobby cartographer and globe maker. He obtained pictures of Luna 3 from the Soviet embassy and used his etching skills to handcraft a Moon globe in 1959, featuring seven thousand craters in a 1:3 relief [see picture above]. In 1969, he finalized an even more detailed Moon globe, showing both sides of the Moon, with eighteen thousand relief features. He created a negative mold to produce resin Moon globes in series, just in time for the moon landing. His masterpiece was a porcelain relief Moon globe, a limited edition of five hundred copies produced by the Hutschenreuter porcelain factory. In the 1970s, Columbus Verlag used a lunar map by Schlegel to produce a plastic Moon globe.

100

Frank Manning (1903-1977) was a colorful figure from New Orleans. He started off as a bodyguard of the Louisiana Governor and was the inventor of Manning's Tasty Shrimp Fish Lure. He was also a hobby astronomer. From 1953 to 1967, he produced Manning's Moonball, a six-inch soft rubber relief ball depicting lunar craters. The lunar globe came in four colors: Silvery, Harvest, Green Cheese, and Blue Moon. As the far side of the Moon had not been surveyed, Manning covered the empty side of his Moonball with facts and figures, cast in huge relief letters, resembling a Roman monument.

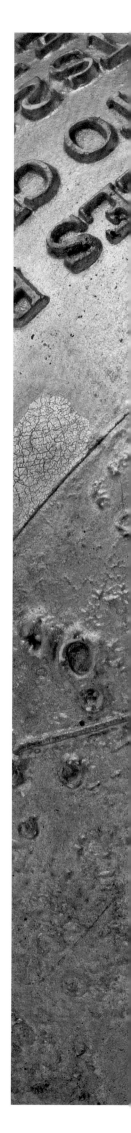

SOVIET MOON

Far Side of the Moon

Nikita Khrushchev (1894-1971) became the new leader of the Soviet Union after winning the power struggle following Stalin's death. He supported the Soviet space program, which successfully led the way to the exploration of the Moon. In 1959, the Soviet Luna 3 was the first lunar probe to fly around the Moon. Its biggest achievement was taking photographs of the Moon's far side, which is invisible from the Earth. It always faces away because the Moon's rotation period matches its orbital period around the Earth. The Sternberg Astronomical Institute in Moscow compiled a lunar map from these images which was then reworked into map gores to create the first accurate lunar globe. This globe was presented at the Moscow Kremlin Theatre in 1959, showcasing the technological progress of the Soviet Union. A year later, a massive lunar relief globe was exhibited at the Leipzig Spring Fair.

**Leipzig
Spring Fair**
A large-size Soviet relief Moon globe is admired by the public /
Leipzig, 1960

Soviets and Moguls

A public version of the Soviet Moon Globe was produced by Räthgloben in Leipzig, East Germany. The globe featured traditional paper map gores adorned with an artistic color wash and shading to depict the lunar geography, creating a visual impression of height and depth. Upon its initial release in 1961, two out of twelve map gores remained blank, as the far side of the Moon had not yet been entirely surveyed [no. 101]. Over the subsequent years, Luna 4 to 8 and Zond 3 captured images of the remaining areas, allowing for the completion of the missing map gores in later editions. Robert Maxwell (1923-1991), a jet-set mogul and owner of the scientific publishing company Pergamon Press in Oxford, secured the distribution rights of the Soviet globe for the Western market. Throughout the following decade, the Soviets produced various lunar globes ranging from 33 to 120 centimeters in diameter.

Lunar Nomenclature

The International Astronomical Union governs the naming conventions for lunar features, permitting only names of scientists and explorers while prohibiting those of political or religious leaders. Korolev, the father of the Soviet space program, achieved some form of vindication with the naming of the Korolev crater, the largest on the far side, while a Stalin crater remains nonexistent. Luna 9 performed a successful first soft landing on the Moon in 1966, transmitting ground-level images back to Earth. The Soviet Union planned a secret program of manned lunar explorations, but after a number of accidents and setbacks, they lost their lead in the Space Race, and Soviet leadership decided to redirect their efforts toward unmanned lunar exploration. In 1970, Luna 16 brought back moon samples, while Luna 17 deployed the first remote-controlled lunar vehicle, the Lunokhod.

101
Räthgloben, the East German globe maker in Leipzig, released this popular version of the Soviet Lunar Globe in 1961. It was a thirty-three-centimeter model, mounted onto a Bakelite base. In the initial edition, two out of twelve paper map gores remained blank, as the far side of the Moon was not yet fully explored. A few years later, the Luna 4 to 8 and the Zond-3 lunar probes captured images of the uncharted regions, enabling Räthgloben to complete its lunar globe. In subsequent editions, all map gores were filled in, depicting the entire lunar landscape.

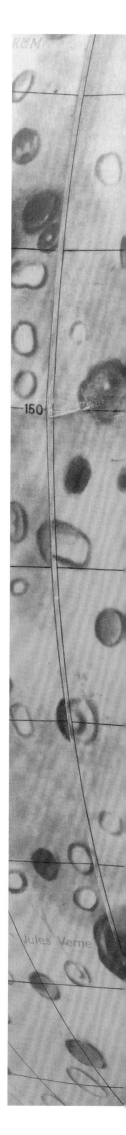

AMERICAN MOON
GLOBES

Nazi Rockets

President John F. Kennedy (1917-1963), announced to enter the Space Race, a few weeks after Soviet cosmonaut Yuri Gagarin became the first human to orbit the Earth, by uttering his famous words *"... the United States should commit itself to achieving the goal, before this decade is out, of landing a man on the Moon and returning him safely to the Earth...."* The endeavor was led by Wernher von Braun (1912-1977), a German rocket engineer who developed the first jet propelled V2-missiles in the 1930s, used by Nazi Germany to bomb London. At the end of the war, von Braun was captured by the U.S. Army and brought to the United States, where he assumed a pivotal role in pioneering the American Moon project at NASA. He designed the powerful three-stage Saturn V rocket for the Apollo missions, which successfully transported twelve astronauts to walk the Moon between 1969 and 1972.

Race to the Moon

The United States initially fell behind in lunar cartography, as many launches of the Ranger moon probes ended in failure. To find the most suitable landing spots, Wernher von Braun had to procure a lunar map from the Soviet Academy. It was not until the Lunar Orbiters of 1966-67 that NASA managed to compile a Photographic Moon Atlas. Due to the lack of lunar cartography and limited public interest, American globe makers seriously trailed behind. In 1958, globe maker Stuart L. Hammond commented: *"There is no real demand for Moon maps yet, but we want to be ready when customers ask for".* Ten years later, the Apollo 11 landing sparked a widespread fascination with the Moon. Globe makers such as Replogle, Cram, Rand McNally, Denoyer-Geppert and Weber Costello all responded by introducing Moon globes based on NASA's photo images, including those of the far side [nos. 102, 109, 111, 167].

Moon Mania

The Moon craze captivated both sides of the Atlantic. Most Moon globes consisted of twelve inch spheres, featuring ocher-colored paper map gores. The cartography offered an artistic interpretation, depicting shaded craters and landing sites. Other models included an inflatable moon by Hammond, plastic moons by Rico and Columbus, metal moons by Chein, Ohio Art, Scan-Globe, MS-toys, and Vulli [nos. 33, 43, 44, 134, 135, 136], and cast-resin relief Moons by Wightman [no. 160] and Nystrom [no. 110]. The president of the American Toy Manufacturers cleverly remarked: *"The Moon trip has turned the toy industry into a branch of NASA".* Rand McNally alone produced one hundred twenty-five thousand lunar globes. However, the hype was short-lived: public interest faded after the last Apollo mission in 1972, and Moon globes became as scarce as before.

Jim Lovell (1928-)
Apollo 8 crew member in space suit next to a Denoyer-Geppert Lunar Globe / NASA, 1972

102
In 1969, Denoyer-Geppert, the Chicago globe maker, received a request from the White House to produce lunar globes for President Richard Nixon (1913-1994) to present as gifts to his foreign guests. The company created a high-quality sixteen-inch Moon globe featuring paper map gores affixed onto a bronze gimbal meridian. The cartography was based on photographs captured by the Apollo 10 mission. As part of a publicity campaign, the crew presented the first copy to Nixon. NASA even commissioned official portraits of its astronauts, dressed in space suits and holding a Denoyer-Geppert globe [see picture above].

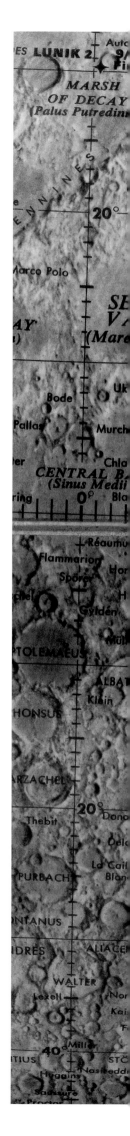

More Celestial Globes

103
Replogle & Scan-Globe published a series of six-inch lithographed tin globes in the 1970s, which included the Earth, the Moon, Mars and a celestial globe. They were sold as a set, and came on various stands, in this case a steel wire tripod base. The celestial map is a scaled-down version of the celestial map that was used for the larger globes.

104
C.S. Hammond & Co., map and globe maker from New Jersey, published a celestial globe in the 1940s. The paper gores of its twelve-inch sphere utilize the star map created by Commander S.E. Stubbs of the Royal Navy, with magnitudes from one to five. The constellations are illustrated both as stick figures and as rectangular zones.

105
Danish Scan-Globe created this twelve-inch constellation globe in the 1970s. The cartography is by Karl F. Harig (1940-). The acrylate sphere displays a blue celestial globe with yellow stars, yet when the light inside is switched on, a colorful depiction of the constellations emerges, with mythological animals floating in a mysterious glow.

More Planetarium Globes

106
The Torica, was a six-inch, funky version of Lafayette's Astro Globe, manufactured in Japan around 1960. Entirely crafted from plastic, it featured a transparent celestial sphere with molded stars on its surface and a turquoise Earth enclosed within. All parts were movable, allowing users to navigate the universe like real astronauts.

107
Amateur astrologer Armand Spitz from Philadelphia designed his Spitz Junior Planetarium in the 1950s. The device consists of a plastic globe with an internal electric lamp, positioned atop a plastic base. Light emitted through three hundred pinholes in the globe's surface creates a projection of the starry sky on the ceiling.

108
The Trippensee company in Saginaw, Michigan, has been producing this Tellurian with chains and gears since 1905. It showcases the movements of the Earth, the Sun and the Moon and allows users to explore the secrets of day and night, the four seasons, and the phases of the Moon. This particular model dates from the 1950s and is crafted from Bakelite.

More Lunar Globes

109

Around the time of the moonshot in 1969, all American globe makers produced lunar globes, such as this model from Rand McNally. It is a twelve-inch sphere with paper map gores. The cartography shows shaded craters to enhance the relief effect. The steel wire stand is listed in a Cram's catalog, so this might very well be composed of parts of different globes.

110

The Educational Frontiers Company produced a seven-inch raised relief Moon globe around 1969, which was distributed by Nystrom in Chicago. It is a topographical relief sphere, made from a yellow-ocher-colored type of molded plastic. The craters are protuding above the surface of the Moon. The sphere rests on a plastic cross cradle base.

111

This lunar model was Replogle's contribution to the Space Race in 1969. It has paper map gores on a twelve-inch plastic sphere. The cartography represents an intricate lunar map, complete with shaded craters and landing sites. The circular metal stand has a indentation to accommodate a manual with astronomical explanations.

More Soviet Globes

112

This Satellite Globe was produced in the Soviet Union in 1961 to commemorate Yuri Gagarin's first spaceflight around the Earth. The five-centimeter gold-colored plastic sphere displays the continents along with a network of longitudes and latitudes. A miniature red rocket orbits the Earth on a steel wire. The name and date are cast into the plastic base.

113

The Globus IMP served as a navigation globe integrated into the control panels of all Soviet spacecraft from 1961 to 2002. Central is a five-inch terrestrial globe, which moves within a transparent half dome, motorized by a complex mechanical clockwork. A crosshair indicates the position of the spacecraft above the Earth.

114

Navigation globes of this type were used on ships throughout the twentieth century. This particular model, belonging to the Soviet Navy, features a sixteen-centimeter celestial sphere, with yellow map gores and Cyrillic star names. The sphere is suspended within vertical and horizontal graded metal rings and housed in a metal case.

POP CUL-TURE

On top of the World

The twentieth century saw the rise of popular mass culture. This was made possible by new mass media, such as cinema, radio, and television. The distribution of recorded movies, music, and images to millions of people could turn an actress or a singer in an international star overnight.

A telling example is Marylin Monroe (1926-1962), a successful actress in Hollywood in the 1950s, who became both a popular sex symbol and a tragic figure. Like many stars, her professional career and her private life were condensed into a single iconic image, exposed for public consumption and exploitation. Monroe was married to boxing champion Joe DiMaggio (1914-1999) and writer Arthur Miller (1915-2005). She also had relationships with famous actors such as Frank Sinatra (1915-1998), Marlon Brando (1924-2004), and Yves Montand (1921-1991). Her last public appearance was at the party of President Kennedy (1917-1963), where she sang a sensuous *Happy Birthday*. A few weeks later, she died of an overdose of barbiturates.

In this promotional picture for the 1952s movie *We're not married!* Monroe, dressed in a negligee, presses her bare legs onto a shiny-blue globe, creating a frivolous contrast with her red lips and red fluffy pumps. The ambiguous message is that Monroe is sitting on top of the world.

Globes have always been used for trivial purposes, aside from their primary geographical function. In the twentieth century, globes were transformed into mini-bars, gadgets, or displays. At the same time, globes became children's toys, sold as games, puzzles, or sharpeners. The many guises of the globe reflect numerous stories about the development of popular culture in the twentieth century.

GAME

Kindergarten

Friedrich Fröbel (1782-1852), a German pedagogue, was the inventor of the Kindergarten. He developed education for children and believed that *"play is the highest expression of human development in childhood for it alone is the free expression of what is in the child's soul."* His idea, novel at the time, was that children constitute a separate group that needs specific clothing, food, education and toys. In the twentieth century, globes followed this evolution: parallel to globes as serious scientific instruments for adults, playful educational globes for children were created. The wave came from two directions. On one side, school suppliers developed school globes with simplified map images, adapted to the cognitive level of children. On the other side, the tin toy factories developed recreational globes aimed at children, which combined play and education.

Baby Boom

After the war, the baby boom opened up a large market for children's products, including globes. The biggest European children's globe makers were Chad Valley, Michael Seidel, and Jouets Mont-Blanc. In the U.S., J. Chein and Ohio Art were prominent companies. Adult globe makers also tried to get a chunk of the new market, including Replogle, which launched colorful metal game globes, designed by chief cartographer Gustav Karl Brueckmann (1886-1962). Larger than the usual toy globes, the eight-inch globes provided more surface for games as well as for geographical information. The idea was to gain geographical knowledge while playing. Replogle wrote: *"... Every thoughtful parent realizes the great value of a globe for the child. When that globe also provides fun for the child, the combination is irresistible. In two years, these new educational toys have won outstanding consumer acceptance...."*

Learn and Play

Replogle's Magnetic-Globe-Race teaches children about foreign countries and remote places. Its spinner base indicates the number of miles players may fly their magnetic airplane pawn across the globe [no. 133]. The Globe-Grams World Game, a sort of geographical Scrabble, combines grammar, geography, and fun. The players spin a metal terrestrial globe that points to letters on its horizon ring. The goal is to compose names of countries, oceans, or cities [no. 132]. The Blast-Off-Globe was advertised by Replogle as *"A space game of fact and fantasy"*. Its black metal globe has the solar system drawn on it: nine colorful planets and the Sun, surrounded by mythological figures, astronauts, and spaceships. The spinner on the base indicates the number of trips each player can make between planets with a magnetic rocket pawn.

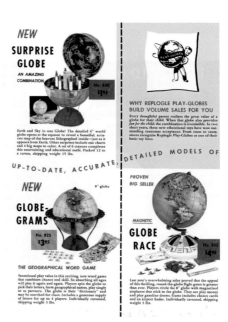

Get on Top of the World!
Replogle Play-Time Geography /
Advertisement for Game Globes /
Chicago, 1954

115

Replogle focused on the postwar children's market: *"A child learns more by living than he does in school. He learns more by touching and spinning a globe than by looking at a map. The globe ... brings alive the places where people of history lived, where wars were fought, books were written and songs were made ... every spin of the globe is inspiring"*. It created this ten-inch Wonder World Globe in the 1940s, animated with airplanes, penguins, steamships, and various other figures. The metal base functions as a casing for the book *The World is Yours*, which contains geography facts and quiz questions.

THE WORLD IS YOURS   REPLOGLE GLOBES

PUZZLE

Dissected Maps

British cartographer John Spilsbury (1739-1769) pasted a map of Europe onto a wooden board and sawed it into pieces. He called it Dissected Maps, the first jigsaw puzzle. The idea was to help students learn about geography in a playful way by fitting the pieces together. Turning a flat map into a geographical puzzle is easy, but turning a globe into a 3D puzzle is harder. Few puzzle globes have been produced over time due to the structural complexity. Faustino Paluzie (1833-1901), a publisher of teaching materials in Barcelona, designed a remarkable version in the late nineteenth century. Twenty-two separate triangular wooden parts form discs, and together the discs form a globe. The inner surface of the building blocks is covered with text and pictures, providing additional space for information, such as the story of the continents, each represented on a separate disc.

Puzzle Ball

Otto Maier (1852-1925) founded Ravensburger in the 1880s, a puzzle company that aspired to mix education with entertainment. Its first successful board game was A Journey Around the World, based on the eponymous book by Jules Verne. In 1891, Maier created a wooden Geographical Puzzle in the shape of the map of Europe. It taught children about geography and developed their eye-hand coordination, shape recognition, and fine motor skills. Learning by doing was Ravensburger's motto, and the company became the world leader of puzzles and board games. At the end of the twentieth century, it patented the Puzzle Ball, a spherical puzzle of five hundred forty pieces that forms a terrestrial globe once assembled. The completed sphere can be inserted in a steel wire stand, allowing the puzzle globe to spin like a real globe [no. 129].

Magic Sphere

Hungarian architect Ernö Rubik (1944-) had an interest in playful mathematical morphologies. In the early 1980s, he patented the Rubik's Cube, a three-dimensional plastic puzzle cube composed of twenty-six separate colored cubes that can be turned in various ways. The aim is to manipulate the cube until all sides display a single color. The Hungarian State Trading Company Konsumex marketed Rubik's Cube and sold three hundred fifty million copies worldwide, making it one of the biggest toy successes ever. Ferenc Molnár, another Hungarian, invented a comparable manipulation puzzle called Puzzle Globe. Instead of a cube, this puzzle is a metal sphere, with the world map lithographed on it, consisting of twenty-six pieces, each of which can be manipulated separately. Players have to keep turning the blocks until the globe is back in the geographical correct order [no. 128].

Victory Geographical Puzzle
George Philip & Son / *Made in England, 1962*

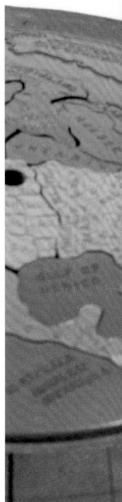

116

The Geographic Educator Corporation in New York made a six-inch puzzle globe in 1927. The World in Pieces, a patent by Charles B. Roberts, can be disassembled into six metal discs, each containing a map puzzle of a continent. The globe was made of a colored resin, developed by chemist Felix Lauter (1882-1942), a pioneer in synthetic materials. The complex globe was expensive, selling at $7.50. After the company went bankrupt during the 1929 Wall Street Crash, the puzzle globes were sold at $2.50, with the slogan: *"On the outside, it looks like the same old world, but when you go inside, the fun begins!"*

PICTORIAL

Mickey and Pluto

Young illustrator Walt Disney (1901-1966) created Mickey Mouse, a twentieth-century icon, after setting up a Hollywood drawing studio in the 1920s. Disney produced animated feature films accompanied by a soundtrack. His comic characters grew so popular that he was able to build a global empire around them, culminating in the creation of Disney amusement parks. The image rights to the comic characters were turned into valuable assets by licensing their use to third parties, such as globe makers. Mickey and his friends continued the historical tradition of pictorial globes, animated with sea monsters and other fantasy animals that sparked imaginations about faraway places. In the twentieth century, the monsters on the globes were replaced by comic characters and happy animals to captivate the imagination of children, the new target group for globes.

Character Licensing

In the interwar period, Ohio Art Company was one of the pioneers in buying licenses for Disney characters to decorate its tin toys. After the war, American globe manufacturers jumped on the new baby boom market with pictorial children's globes, covered with colorful characters from popular children's media. Chicago's Rand McNally, for instance, created a tin globe with Disney characters. Aside from the regular geographical information, popular characters such as Mickey Mouse, Donald Duck, Goofy or Pluto were depicted. The latter was named after the ninth planet, which was discovered a decade before. The Walt Disney World Globe was promoted with the slogan: *"Find out about the world with us - it's fun!"* In addition, Disney published a game globe that depicted the Disneyland attractions, a souvenir of Disney's amusement park, which had opened in Anaheim, California.

Elephants and Poodles

Cram commissioned cartoonist Dave Gerard (1909-2003) to create a pictorial Toy Globe for children in an attempt to compete with Disney's globe. Its map image is simplified to its bare basics, simply showing white landmasses in a light-blue ocean. Its main features are large colorful animals, each one drawn on the continent where it belongs: a penguin on Antarctica, an elephant in India, a whale in the Ocean and so on. Ironically, the U.S. is represented by a poodle, and Europe by a Saint Bernard dog carrying a schnapps barrel [no. 130]. In the late twentieth century, television became a popular medium for children and a great supplier of comic characters. Rand McNally created a Sesame Street World Globe in the 1980s, featuring images of puppets from the popular *Sesame Street* TV show, which still runs in many countries today, even after fifty years.

Mickey Mouse
Two black circles on a globe are enough to recognize the twentieth-century icon

117
George Philip & Son in London created the Philip's Pictorial Globe in the 1950s. The drawings on the ten-inch globe resemble the work of cartographer Ernest Dudley Chase (1878-1966), famous for his pictorial maps. This genre, popular in the U.S., added drawings of economic activities, animals, monuments, or traditionally dressed people into the cartography. Its purpose was to illustrate local cultures using a narrative of twentieth-century stereotypes. The style links back to the pictorial tradition of European maps and globes of the 1500s, with their illustrations of foreign wonders.

GADGET

Servants to Human Needs

Peter Reyner Banham (1922-1988) was a British architectural critic. In one of his articles, he formulated a lengthy definition of the American gadget: *"A characteristic class of U.S. products ... is a small self-contained unit of high performance in relation to its size and cost, whose function is to transform some undifferentiated set of circumstances to a condition nearer human desires. The minimum of skills is required in its installation and use, and it is independent of any physical or social infrastructure beyond that by which it may be ordered from catalog and delivered to its prospective user. A class of servants to human needs, these clip-on devices, these portable gadgets, have colored American thought and action far more deeply than is commonly understood"*. Gadgets have been omnipresent in the popular culture of the twentieth century, and a number of them were disguised as globes.

Ink Wells and Ball Pens

One notable gadget from the 1930s is a Bakelite inkwell in the form of a globe. It was produced in various colors by Sanford, the largest U.S. ink producer. Inkwells were omnipresent in the twentieth century for refilling fountain pens, until one day in the late 1930s when Hungarian journalist László Bíró (1899-1985) stained his hands while filling his fountain pen. He was so irritated that he invented the ballpoint pen or Biro, a pen with a rolling metal tip that does not drip. During the Second World War, the Jewish Bíró was forced to flee Europe and seek refuge in Argentina. He sold his invention to BIC, a French company that proliferated the plastic ball pen as the most successful writing tool of the twentieth century. In 1947, the New York Daily News called the ball pen *"The Outstanding Gadget of the Postwar Era"*. Sanford countered the decline of the fountain pen by inventing the Sharpie, the felt-tip pen.

Sharpeners

Another traditional writing tool was the pencil, a graphite stick encased in wood. Pencils require regular sharpening, so sharpeners in many forms and shapes have been invented, often as gadgets. At the start of the twentieth century, German tin toymakers build sharpeners into their miniature globes, making use of the hollow space as a grinding container. Although small in size, the cartography on these early tin globe sharpeners is rich and elegant, thanks to sharp lithography, bright colors, and curvy graphics of ocean currents. Some featured a cast-iron sculpture as a stand, such as the muscular Atlas figure, who carries a grind-filled globe on his neck [no. 126]. In the 1950s, KUM, a West German pencil sharpener factory, made a small, lithographed tin sphere attached to a cast-iron grinding mechanism, operated by a wooden mill handle.

Globe-sharpener packaging box
Color-printed cardboard depicting inside and outside / *Made in Hong Kong, 1960s*

118
Shortly before being purchased by Weber Costello in 1900, American Globe and School Supply in Seneca Falls, New York, produced a series of paperweight globes, in collaboration with Rand McNally in Chicago. The three-inch miniature model, designed to serve as a paperweight on a study desk, was available with either a wooden ball or a magnifying glass stand. These globes allowed users to keep their documents organized, while also providing a way to locate places on Earth. Paperweights gradually disappeared in the late twentieth century as the shift to computers and the digitalized paperless office reduced wasteful paper stacks.

KITSCH

Alcohol

British secret agent James Bond is a notorious womanizer and martini sipper. His famous line to any bar tender is: *"Shaken, not stirred"*. Celebrated in movies, alcohol and tobacco remained important intoxicants in the twentieth century, despite prohibitions and taxations. There is a long tradition of globes being used for the sake of tobacco or alcohol. Rummer-Globes already existed in the sixteenth century. They had hemispheres crafted from precious metal, designed to separate as and form two rummers for enjoying wine. A contemporary variation is the Bar-Globe, a popular kitsch object found in gentlemen's studies. The large globe, usually a floor model designed with questionable taste, is equipped with hinged hemispheres that open to reveal an interior filled with liquor bottles, attracting more of the guest's attention than the cartography [no. 119].

Tobacco

Parisian globe maker J. Forest offered a Lighter Globe for gentlemen around 1900. The small metal sphere, covered in miniature paper map gores with detailed cartography, was actually a gasoline holder with a fire tube located at the North Pole [no. 127]. It was considered uncivilized for women to smoke, so Forest offered a similar model for ladies, the Perfume-Globe, with a spray nozzle attached to the North Pole. Fifty years later, smoking habits had changed: after the war, smoking was considered a virtue for both men and women. Social etiquette demanded that the lady of the house elegantly present her guests with various brands of cigarettes. To facilitate this, Japanese lighter company Windmill invented the Cigarette-Globe, a popular kitsch object made of chrome and brass. When lifting the northern hemisphere, a bouquet of cigarettes would spring up.

Sweets

British biscuit factory Huntley & Palmers pioneered tin jar packing. It helped to preserve and ship biscuits safely to faraway places, allowing the company to expand into one of the world's largest biscuit factories. The lithographed tin biscuit jar also was a perfect marketing tool: the firm created elaborate models, decorated with various themes, colors, and shapes. One model from the early 1900s is a seven-inch tin globe with a political map lithographed on it. The Biscuit-Globe rests on four ball feet. The top has a hinged lid that reveals an array of sweets inside the globe. William Crawford & Sons, a competing British biscuit factory, offered its own tin Biscuit-Globe in the 1930s. It was a Reliable Series Globe by toymaker Chad Valley, but in this case with a hinged hemisphere to open it, produced by the tin packaging factory of Barringer, Wallis, and Manners.

A Satisfied Customer Advertisement for Marlboro cigarettes using a smoking globe / *1960s*

119
The Dutch distiller Van der Tuin brewed a local brandy as an alternative to imported French cognac. To give it a French feel, they translated their family name into the French Dujardin. Around 1960, the brand promoted the faux cognac by means of a Bar Globe. It was a thirty-four-centimeter standard model by German globe maker Columbus Verlag, adapted with a hinged meridian so the top hemisphere could be opened. Inside the globe, there was a lush gold-padded interior that held a bottle of Dujardin and six glasses. The hinged model was patented in 1957 by Paul Oestergaard, owner of Columbus Verlag.

"A globe at their own ... makes homework easier ... helps them get better marks ... inspires them to become good citizens ... To touch Rome or London, to put a finger on a mountain – The gift that inspires children to learn ... their own world to explore!"

Luther Replogle, 1955
Owner of Replogle Globes
Chicago globe maker

A group of Sami children having fun finding their northern position on a globe / *1936*

CHEIN TOY

GLOBES

Junk Food

Jewish-Russian Julius Chein (1873-1926) emigrated to the U.S. with only one arm, having lost the other in a fireworks accident. He founded J. Chein & Co., a metal workshop in Manhattan in 1903, where he produced cheap tin toys, sold at Five and Dime stores or as prizes for Cracker Jack's, the first sugary junk food. During the 1910s, Chein registered patents for mechanical toys and built a factory in Harrison, New Jersey. The import of German tin toys halted during the First World War, and Chein filled the void with his assortment of colorful lithographed toys equipped with mechanical springs, ranging from tambourines to trains and merry-go-rounds. Chein was a big brand in the golden age of tin toys. Julius became a rich man, known for his temper: once, triggered by a problem in the factory, he threw his watch on the floor and furiously jumped on it.

Tail Fins

Julius Chein died of an epileptic attack during a horse ride in Central Park. His widow passed the helm of the company to her brother, Samuel Hoffman (1890-1976). Hoffman started his career as a vaudeville artist before founding his own toy company, the Mohawk Metal Toy Co. He went on to further expand Chein's toy assortment, adding metal bank globes with a coin slot, as well as advertisement globes with a customized logo at the base [nos. 86, 120]. During the war, production was interrupted, and the company supported the war effort by manufacturing metal tailfins for bombs. Paradoxically, Chein experienced a lot of competition from defeated Japan in the 1950s. As part of the postwar agreement, the U.S. forced Japan to focus on low-yield products. As a consequence, cheap 'Made in Japan' toys flooded the American market.

Safety Regulations

Sons-in-law Irving Sachs (1918-2012) and Robert Beckelman (1925-2020) took over after the war. They moved to a larger plant in Burlington, New Jersey, employing six hundred workers and producing thousands of toys daily, including terrestrial globes and Moon globes [no. 33]. In the 1960s, Chein encountered fierce competition from plastic toys that could be produced faster and cheaper. In the 1970s, consumer safety rules created even more challenges: sharp-edged toys were prohibited, leading to the closure of Chein's metal toy department. The company was renamed Cheinco Industries and refocused on household items, such as food cans and trash buckets. Due to its polluting factories, Cheinco faced further difficulties when environmental regulations tightened, resulting in bankruptcy in 1992. The remaining assets were taken over by the U.S. Can Company.

Chein Tin Toy Catalog
Colorful spinners and globes /
Burlington, New Jersey, 1959

120

J. Chein & Co. from New Jersey specialized in colorful globes made of lithographed metal. These small globes consisted of two pressed metal hemispheres, clicked together at the equator. This particular model from the 1930s is more sophisticated. The three-and-a-half-inch sphere is mounted on a metal meridian to allows it to spin, and the cartography is more detailed, enhancing its educational value. A circular metal disc forms the base, often customized with brand names or decorative prints. In this case, it features a calendar with the months, the seasons, and the names of the zodiac signs.

OHIO ART TOY
GLOBES

Picture Frames

Dentist Henry S. Winzeler (1876-1939) sold metal picture frames in Archbold, Ohio. In 1910, he bought stamping machines and began producing his own frames. The Ohio Art Company soon expanded, making twenty thousand frames a day, and moved to a larger plant in Bryan, Ohio. During the First World War, the import of German tin toys halted, and Ohio Art seized the opportunity to enter the market. It developed a range of tin toys, such as sand buckets, trains, and spinning tops. To make the toys more attractive, artist Fern Bisel Peat (1893-1971), a well-known illustrator of children's books, was commissioned to design colorful decorations. Ohio Art also acquired licenses for Disney characters. After Winzeler's retirement, his successor Lachlan M. MacDonald (1886-1967) steered the company through the Depression by broadening the range of products, from car parts to tumble dryers.

New Deal

President Franklin D. Roosevelt launched the New Deal in the 1930s. His economic recovery plan was a combination of fair competition, fair prices, and worker protection, such as minimal wages and maximum working hours. Ohio Art signed up and recruited a hundred new staff. In the next four decades, the company produced many children's globes. The four-inch metal spheres were fixed on a circular metal base, decorated with the zodiac, animal figures, or airline logos by Kermit Bishop (1904-1989) [no. 14]. Bishop was a self-taught artist and cartographer who had worked earlier for Rand McNally. The larger six to ten-inch metal globes featured a more elaborate cartography and were equipped with a meridian that allowed them to spin, effectively competing with paper globes [no. 121]. Even a ten-inch metal Moon globe was available [no. 136].

Made in China

Howard W. Winzeler (1915-1995), son of the founder, led the company through the 1950s. He made toy history when he bought the rights to Etch-A-Sketch, a glass screen with aluminum powder that can be drawn on by turning two knobs. Ohio Art sold ten million copies in the 1960s alone. During the same period, it purchased a plastic injection molding company to diversify to plastic toys. The company also made metal products on demand, such as Coca-Cola trays or Kodak film canisters. In 1966, William C. Killgallon (1913-2000) became president and bought the company a few years later. Like many industries, Ohio Art relocated its production to China. About eighty percent of toys sold in the U.S. today are made in China, often under poor labor conditions, such as an eighty-four-hour week.

Fascinating Family Fun!
Advertisement for Etch-A-Sketch / Ohio Art Company / Bryan, Ohio, 1960s

121
The Ohio Art Company in Bryan, Ohio, made this colorful tin toy globe in the 1950s. Kermit Bishop, the company's globe designer, included a lot of information on this small object. He used the image of the Greek god Atlas as a flat metal figure to carry the six-inch tin lithographed sphere. The cartography is detailed, with different colors for each U.S. state. The base is decorated with the names of the months, the seasons, and zodiac signs. The globe is more than a toy; it is an educational object that offers children an introduction to geography and sparks curiosity about the world.

CHAD VALLEY TOY
GLOBES

Teddy Bears

Joseph (1842-1904) and Alfred Johnson (1846-1932) continued their father's printing business in Birmingham in the 1860s. The firm specialized in office stationery and board games, made of printed card. Their board games became more sophisticated, sales increased, and the brothers constructed a new factory in Harborne, in the Valley of the Chad Brook, leading to the company's brand name: Chad Valley. Around the turn of the century, a handicapped seamstress, German Margarete Steiff (1847-1909), created a stuffed toy bear and exported thousands of copies to the U.S. The bears were nicknamed Teddy Bear, after President Teddy Roosevelt (1858-1919), who once refused to shoot a young bear on a hunt. Chad Valley followed the trend and expanded its range of board games to include soft toys and stuffed dolls, products that made the company famous.

Toymakers to the Queen

Alfred J. Johnson (1873-1936) took over the company from his father and uncle and expanded Chad Valley into the dominant force in the British toy industry of the twentieth century. The company introduced many innovations, such as celluloid doll heads to replace the traditional heads made of fragile porcelain. Just before the war, Chad Valley obtained the Royal Warrant of Appointment: *Toymakers to Her Majesty the Queen*, after creating dolls that resembled the young princesses. The company had an expansionist policy, to diversify production, it purchased other specialized factories, such as those for Bakelite or tinplate. Barringer, Wallis, and Manners, a specialist in decorated tin packaging, merged into the Metal Box Company and manufactured Chad Valley's tin toys. Over forty years, Chad Valley produced small tin globes with lithographed spheres and stands [nos. 12, 122].

Economic Crisis

Chad Valley's range of globe models was limited. Their common model was sold under the brand name Reliable Series. It featured a folded metal meridian and detailed cartography lithographed onto the tin sphere. In the 1960s, the company introduced plastic parts, used for instance for the meridian, but it did not transition to full plastic globes. During the moonshot period it produced a tin lunar globe with a blank far sideside [no. 134]. Chad Valley was a leading British toymaker; in the 1960s, it employed over a thousand people. In the 1970s, the U.K. was hit by a severe economic crisis, marked by unemployment, inflation, and strikes. Sales of high-end toys decreased, and the company had to downsize. It was bought by Palitoy in 1978, overtaken by Woolworths ten years later, and today it is owned by Sainsbury. Globe production ended a long time ago.

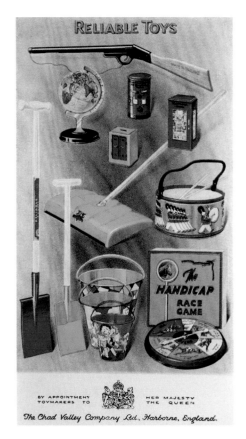

Reliable Toys
Advertisement for tin toys by The Chad Valley Company / Harborne U.K., 1950s

122
Chad Valley in Birmingham, U.K., was one of the biggest British toy factories of the twentieth century. This joyful tin toy globe dates from the 1930s. The globe seems inspired by the constructivist compositions of Russian designer El Lissitzky (1890-1941). The turquoise seven-inch tin sphere spins on the tip of a stark-red triangle of folded metal, which stands upright on a red circular base. To complete the design, a red ball atop the North Pole suggests the Moon circling around the Earth. The small toy globe was a good export product; this example is an Italian-language model.

MICHAEL SEIDEL TOY
GLOBES

Volkswagens

Michael Seidel (1858-1932) founded a tin toy factory in Zirndorf, close to Nuremberg, the German toy capital, in 1881. The factory produced metal toys such as humming tops. German tin toys evolved from handcrafted and hand-painted objects to color-lithographed, machine-folded mass products, exported worldwide. His son Georg Seidel (1884-1950), took over the company, expanded the assortment, and moved to a larger site in the early twentieth century. The Capricorn, his father's zodiac sign, became the new logo. One of the company's successful products was a money box, which sold 4,5 million copies. The German government ordered a customized money box for a national campaign called 'Kraft Durch Freude', to encourage families to save for a Volkswagen. During the war, Seidel was forced to help the war economy by making tracer ammunition.

Metal and Plastic

After the war, the sons-in-law Gerhard Hamann (1902-1987) and Karl-Heinz Müller (1914-1989), resumed toy production and expanded the range of tin toys with cars, trucks, planes, and also tin globes. Seidel responded swiftly to the new materials in the 1950s and bought a plastic injection molding machine to produce plastic toys. The company's catalog contained more than one hundred and fifty colorful products, often combining tin and plastic. The cheerful lithographed tin spheres were mounted on brightly colored, organically shaped plastic stands, inspired by the design of the period [no. 123]. One model has a white S-shaped base that ends in a meridian, creating the impression that the globe is riding a big wave. From the 1960s onwards, the company also made larger globes with twenty-centimeter illuminated plastic spheres, mounted on a metal base [no. 158].

Capricorns and the Climbing Club

In the 1970s, the company changed its name to MS-Toys and restyled the Capricorn logo. Many different languages can be found on Seidel's globes, as they were exported to numerous countries. Their assortment included an inflatable vinyl globe and a metal Moon globe, produced around the time of the moonshot [no. 44]. In 1982, globe production at MS-Toys ended, when the family sold the business to Martin Fuchs FMZ, another toy family in Zirndorf. Home to the Playmobil factory, Zirndorf still remains a toy town, although the transformation from a manufacturing economy to a leisure economy in the late twentieth century is well visible in the town: the former MS-Toy factory building has been refurbished into an indoor climbing club, called Der Steinbock (The Capricorn), a nod to the old factory logo that still adorns the facade.

Für meinem Wagen
Money Box to save coins for a Volkswagen car /
Michael Seidel /
Germany, 1930s

123
Michael Seidel's factory made cheerful toy globes of tin and plastic in the small town of Zirndorf, Bavaria. The model above, dating from the 1950s, is unusual for a European globe, because the fifteen-centimeter tin sphere is lithographed with black oceans. As it is an English-language globe, it was probably meant for export to the U.S., where black globes were more fashionable at the time. The combination of the black oceans, crisscrossed by white ships and airplanes, resting on a stark-red plastic base in the form of a wave, lends a striking effect to the globe.

JOUETS MONT-BLANC

Postwar Prosperity

Joseph Vullierme (1911-2005) was born in Rumilly, a small town at the foot of the French Alps. He learned blacksmith skills from his father and fine mechanics at l'École Nationale d'Horlogerie. After serving as a resistance fighter during the Second World War, Vullierme returned to Rumilly and founded a metal company. It took off when Vuillerme made a successful miniature tin milk wagon as a promotional gadget for the local dairy factory Lait Mont-Blanc. Combining his blacksmith craftsmanship and watchmaker skills, he created an empire of mechanical toys, catering to the children of the baby boom generation. The factory produced Picotin the Windable Chick, the Mont-Blanc Express cogwheel train, and many more, under the trade mark Jouets Mont-Blanc, abbreviated to JMB. By the mid-1950s, it was the largest company in Rumilly after the milk factory, employing over one hundred people.

Metal to Plastic

The Jouets Mont-Blanc catalog contained lithographed tin metal globes with a diameter of twelve centimeters, designed as children's globes for home or school use. The surge of plastics in the postwar era unleashed a revolution in the toy industry. Plastics are inexpensive, strong, and can be molded into any possible shape or color. JMB made a gradual transition from metal to plastic. In the 1950s, the globes were still made of metal; in the 1960s, the base and the meridian were plastic; and finally in the 1970s, the globes were entirely made of plastic. The company also produced illuminated globes made of semi-transparent acrylics. To accommodate a lamp, the spheres were made larger, up to twenty-five centimeters, which allowed for more detailed cartography. These globes had greater educational value, allowing them to compete with traditional paper map globes.

Brand Value of a Rubber Giraffe

In the twentieth century, a brand name was a valuable asset for a company. The German fountain pen factory Mont-Blanc claimed to have registered its brand name earlier than Jouets Mont-Blanc. Vuillerme lost the case and changed the company's name to Creations Vuillerme, shortened to Vulli. During the moonshot period, the company made a metal Moon globe that came with magnetic rocket pawns to mark the landing sites of the Soviet and American spacecraft [no. 135]. To diversify, Vuillerme merged in the 1980s under the brand name Vulli with Delacoste from Asnières-sur-Oise, a factory of rubber toys. Their most successful product was Sophie la Girafe, a rubber figure which sold more than fifty million copies. Finally, in 1989, Vulli was taken over by the Alain Thirion Group. The factory is still located in Rumilly today, but it no longer manufactures globes.

JMB Toy Catalog
Plastic and tin
toy collection /
Jouets Mont-Blanc /
*Rumilly, France,
1960s*

124

Vulli, based in Rumilly in the French Alps, initially made colorful metal globes aimed at children but gradually shifted to plastic globes. This illuminated globe from the 1960s is a full plastic model. Its eighteen-centimeter sphere is made of plastic, with French-language cartography printed on it. The cartouche reads *"Vuillerme S.A., Rumilly - Printed in Italy"*, which might indicate that the plastic sphere was made by Rico in Florence, an early master of plastic globe production. The sphere is fixed to a large transparent meridian and an organically smooth shaped black plastic base.

More Gadget Globes

125

French tin toy factory Jouets TMF in Mulhouse produced this joyful object in the 1950s. The twelve-centimeter lithographed tin sphere with a coin slot can be used as a money bank, while the steel wire stand integrates holders for a pen, color pencils, and business cards. The colorful cartography and the golden wire on red tips make it a playful object.

126

A German tin toy maker built a sharpener into a five-centimeter tin sphere in the early 1900s. It makes use of the hollow space as a grinding container and is carried by a miniature cast-iron Atlas sculpture. Although the size is small, the cartography is rich and elegant, thanks to sharp lithography, bright colors, and curvy graphics of ocean currents.

127

Parisian globe maker J. Forest offered a Lighter Globe around 1900. This gadget and conversation piece for gentlemen consists of a small three-inch bronze sphere covered in miniature paper map gores that show detailed cartography. The sphere is, in fact, a disguised gasoline holder, with a fire tube fixed at the North Pole to ignite it.

More Toy Globes

128

Hungarian Ferenc Molnár invented a manipulation puzzle in the 1980s called the Puzzle Globe. It consists of an eight-centimeter metal sphere, with the world map lithographed on it, comprising of twenty-six pieces interconnected inside by a complex system of hinges. Each piece can be rotated until the globe is correctly assembled.

129

The German company Ravensburger is the world leader of puzzles and board games. At the end of the century, it patented the twenty-centimeter Puzzle Ball, a three-dimensional puzzle comprising five hundred and forty pieces. Once assembled, the globe can be inserted into a steel wire stand, allowing it to rotate like the Earth.

130

Globe maker George F. Cram enlisted cartoonist Dave Gerard to craft a seven-inch pictorial globe in the 1950s. Gerard upheld the tradition of featuring animals on the map gores, showcasing a penguin on Antarctica, an elephant in India, and a whale in the Pacific Ocean. The U.S. is represented by a poodle, Europe by a Saint Bernard dog carrying a schnapps barrel.

More Game Globes

131

in the 1950s, Replogle produced a six-inch children's globe known as The Surprise Globe. It comprises two hinged metal hemispheres, featuring a terrestrial map on the exterior. Upon opening, a hollow celestial map is revealed within the globe, accurately depicting the concave universe as observed when looking up at the night sky.

132

Replogle also created the Globe-Grams World Game in the 1950s. This geographical word game is played on an eight-inch metal sphere, combining grammar, geography, and fun. Players spin the terrestrial sphere that points to letters on the horizon ring. They then compile names of countries, oceans, or cities, using letter tokens.

133

Another model in Replogle's 1950s Game Globe series is this Magnetic Air-Race Globe. The globe has a spinner on its base, indicating the number of miles players may fly with their magnetic airplane-shaped pawns on the eight-inch metal sphere. Through gameplay, children learn about foreign countries and remote places.

More Lunar Toy Globes

134

British toy producer Chad Valley jumped onto the Space Race in the 1960s by offering a small four-inch lithographed metal Moon globe on a plastic base. The far side has a dark purple color, an inaccurate reference as both sides receive equal amounts of daylight. Contrary to other toy Moon globes, its cartography is rather basic.

135

In around 1969, French tin toy factory Jouets Mont-Blanc in Rumilly produced a small tin Moon globe on a plastic base. The ten-centimeter metal sphere features a lithographed simplified lunar map. One can follow the Space Race by placing the red and white magnetic spacecraft on the metal Moon globe to indicate the Soviet and American landing sites.

136

The tin toy factory Ohio Art in Bryan, Ohio, created this Moon globe in the 1960s. Featuring a large nine-inch tin sphere, it showcases a meticulously detailed lithographed lunar map, complete with an American flag marking the Apollo 11 landing site. The Moon has an unusual strong-blue hue, contrasting with the faux-wood pattern on the tin base.

EU-ROPE

World Wars

A terrible new phenomenon emerged in the twentieth century: world wars. This type of armed conflict involves many nations and is waged simultaneously across multiple locations worldwide. It is estimated that more than 200 million people lost their lives in wars during the twentieth century.

Wars end in highly mediatized, symbolic moments, such as the tearing of the enemy's flag from a capital building, or the seizing of their globe. In the 1930s, the German Nazi regime commissioned a number of monumental globes for state buildings, such as the Reich's Chancellery, the so-called 'Globus für staatliche und industrielle Führer', made by Columbus Verlag, which consisted of a 106-centimeter aluminum sphere on a heavy wooden stand, with a very detailed cartography. Legend has it that when the Red Army liberated Berlin in 1945, Soviet soldiers shot holes in these globes at the spot where Germany is located. Although there are indeed a number of globes on display in museums with bullet holes and other damage, their authenticity is contested. The fate of these globes remains unclear, fueling public imagination even today. Most likely, these globes were destroyed during the fall of Berlin: in this 1945 photograph, a damaged globe with bullet holes can be seen alongside a broken bust of Hitler, amidst the ruins of the Reich Chancellery in Berlin.

Throughout the twentieth century, European globe makers evolved from small family businesses, each boasting their own history. Some ventured into globe making after producing road maps, others began as publishers of school supplies, and some simply pursued it as a hobby. Despite navigating economic crises and world wars, European globe makers transformed their small workshops into large factories.

TARIDE

Bicycle Mania

Louis Alphonse Taride (1825-1859) established a bookstore under the arches of the Odéon Theatre in Paris in 1852. His first successful publication was *Paris à Pied et à Cheval*, a city map for hikers and horse riders. His son Henri Alphonse Taride (1859-1918) moved the bookstore to the Grands Boulevards at Porte Saint-Denis, the center of intellectual life during the Belle Epoque. At the end of the nineteenth century, the bicycle gained popularity: people went for cycling trips around Paris. Taride's answer to the bicycle mania was the first folded road map of Paris and its surroundings. Its great success propelled the company to become one of the most important twentieth-century French road map makers. Taride also published plans of the Metro, city guides, and a popular cookbook titled *La Cuisine de Tante Marie*, which could be found in every French household.

Import and Export

Grandsons Jean (1910-1996) and Claude Gourier (1920-1974) took the helm after the Second World War. They turned their focus to geographical products and globes. Every year, Taride produced thousands of globes and more than sixty different models. Their selection included cardboard spheres, tin globes, and illuminated glass globes that had the company's logo cast into the electric plug. The smaller models were fixed to simple steel wire legs, the larger ones to more elaborate bent metal stands [nos. 3, 137]. Taride exported many globes and collaborated with foreign publishers like the British George Philip & Son. The company also distributed foreign brands in France, such as Scan-Globe from Denmark and Rico from Italy. In the 1960s, Taride opened a new factory in Issoudun, south of Paris, where the Gourier family had its roots.

Jetset on the Côte d'Azur

The brightly colored postwar Taride globes, with their mischievous legs and ball feet, exude the comically absurd atmosphere of Jacques Tati (1907-1982) movies of the period [no. 157]. An interesting example is a globe commissioned by Taride from designer couple Adrien Audoux (1903-1991) and Frida Minet (1897-1992), famous for their modernist abaca and seagrass furniture. They designed a globe with a twist: an illuminated glass sphere, resting on a fisherman's rope. It appealed to the bohemian jet set at the Côte d'Azur, where the couple had their studio [no. 153]. Taride also created inflatable globes in collaboration with rubber boat company Sevylor [no. 17]. At its peak, the company employed more than fifty people, mainly women who had previously worked at Issoudun's confection ateliers. After forty years of popular French globes, Taride closed its doors in 1986.

Cartes Taride for cyclists and motorists
Promotion poster
Taride / *Paris, 1899*

137

Taride from Paris made attractive globes, renowned for their cartography, colors, graphics, and stands. The above 1960s model is a good example of a modernist atomic style globe. It features a twenty-two-centimeter sphere with paper map gores and was also available as an illuminated glass globe. The cartography boasts a fresh color scheme. The sea currents are depicted as dynamic, curvy white lines that stir up the electric-blue oceans. The stand consists of a black metal strip, bent in a single motion that starts at the circular base and continues into a meridian that holds the sphere at the poles.

GIRARD & BARRÈRE

Parisian Legacies

Joseph Forest (1865-1923) was accredited by the Ministry of Education, which allowed him to sell geographical materials to schools, such as maps and globes. It turned his company into one of the most important French globe makers, with an extensive production of terrestrial as well as celestial globes. Forest successfully exported globes to Canada and Latin America. As a result, the French government appointed him as Foreign Trade Advisor and rewarded him the Legion d'Honneur. Meanwhile, Henry Barrère (1865-1930), a contemporary of Forest's, took over the house Andriveau-Goujon and continued to publish globes both under their as well as his own name. A third Parisian house was Bertaux & Thomas, successor to the famous Maison Delamarche. The three companies produced traditional papier-mâché globes in fin-de-siècle style, set on a cast-iron or wooden stand.

Mergers and Brands

In the 1930s, E. Girard acquired Forest, Barrère, and Bertaux & Thomas. He consolidated the three houses into one new Parisian cartographic company. The company published traditional globes under the Forest brand and more fashionable art deco-styled globes under the name of Girard Barrère. Their spheres were made of papier-mâché or illuminated glass, mounted on geometric stands in folded metal with a pin at the South Pole, or on a metal ring [no. 35]. The cartography was revised with pastel colors and sweeping sea currents. The globes specifically targeted the world savvy family man: *"Own A Modern Terrestrial Globe - You will spend pleasant evenings with your family - You will follow the raids of our airmen around the world - You will be interested in global events - The globe is the only accurate map of the Earth"*.

Atomic Style

After the war, the company changed its brand name to Girard, Barrère & Thomas and abandoned art deco in favor of the more popular 1950s atomic style. The elegant globes float on black steel wire tripods, similar to the way the Citroën DS glided across the roadway, carried by its pneumatic suspension system [no. 138]. The map gores were made of laminated cardboard, which were glued to plastic spheres. They depict light-blue oceans and freshly colored land. Girard, Barrère & Thomas advertised its globes as decorative objects, rather than geographical instruments: *"To furnish your offices and studios, we put at your disposal ... terrestrial globes."* However, the company was late to modernize and missed the transition to plastic globes. In 1974, competitor Taride took over the declining company, only to have to shut its own doors a few years later, thereby ending three hundred years of French globe making.

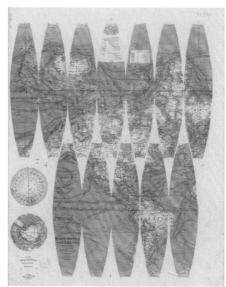

Map gores for a globe
Color-printed cartography on paper sheet /
J. Forest /
Paris, 1920s

138
Girard, Barrère & Thomas in Paris created frivolous globes in the postwar era.
A fine example is this 1960s model featuring a twenty-five-centimeter sphere that seems to float on top of a black steel wire tripod that encompasses the meridian in one continuous dynamic movement.
The lightweight legs are reminiscent of the antennas of the Sputnik satellite.
The cartography features a composition of light-blue oceans and fresh land colors.
The map gores consist of laminated cardboard, affixed to the plastic spheres with contact glue, which unfortunately creates a brownish seam over time.

COLUMBUS
GLOBES

Ein Globus in Jedes Heim!

After working at his brother's publishing house, Paul Oestergaard (1874-1956), a Danish teacher in Berlin, founded Columbus Verlag in 1909. It was his ambition to make a 'Volksglobus', affordable to everyone: *"Like the Radio and the Newspaper, a Columbus Globe in Every Home!"*. Oestergaard bought a steam press to form smooth machine-made papier-mâché spheres and mounted the map gores directly on them. It allowed him to produce larger quantities in a shorter time, for a lower price. Around the First World War, Columbus Verlag already exported globes in twenty-four languages, including all the way to China. During the interwar period, the company created both popular globes and luxury globes, such as the Globus für staatliche und industrielle Führer, a 106-centimeter aluminum sphere with detailed cartography on a monumental hardwood stand.

The Berlin Blockade

Paul Oestergaard Jr. (1904-1975) succeeded his father during the war. The family survived by making globes at home, since its factory had been destroyed by the Allied bombing. After the war, the Soviet Union imposed a blockade of the western sector of Berlin in 1948, marking the start of the Cold War. The Oestergaard family decided to move to West Germany and rebuilt their life in Stuttgart. Columbus produced some of the most precious globes of the twentieth century. It offered eighty-four different models, including celestial, lunar, illuminated, and relief globes [nos. 29, 36, 50, 55, 70, 79, 80, 93, 119]. It devised several innovations, including the Duo Leucht-Globus, that can display both a political and a physical map [nos. 149, 152]. Grandson Peter Oestergaard (1936-) initiated the plastic globe production in the 1960s with the Duplex Globus: a map printed directly onto a flat plastic disc, which is then vacuum pressed to form a hemisphere.

Collapse of the Soviet Union

In the 1990s, the Soviet Union and Yugoslavia collapsed, making the permanent adjustment of maps necessary. The market deteriorated and production moved to low-wage countries in Asia. Personnel at Columbus shrunk from one hundred and thirty to eight collaborators, and the family considered closing. Great-grandson Torsten (1966-) decided to rebuild from scratch once more. He moved to a smaller workshop and introduced digital cartography. The situation improved, and Columbus was able to take over the staff of the bankrupt Räthgloben. It continued to innovate by introducing the first levitation globe at the turn of the millennium [no. 187]. Today, it is one of the few remaining European globe factories, with exports amounting to half of its yearly production. Since its inception, Columbus Verlag has sold more than ten million globes, fulfilling its founder's ambition, *"A globe in every house!"*.

The Earth in Words and Numbers
Manual for a terrestrial globe / Columbus Verlag / Berlin, 1930s

139
In the 1950s, Columbus Verlag in Stuttgart created this luxury floor globe that is still in production today. It is called the Adenauer Globe, in reference to the first West German Chancellor, Konrad Adenauer (1876-1967), who had a copy in his office. The large floor globe includes a fifty-centimeter illuminated glass sphere, encircled by a wishbone-shaped walnut stand. The sphere is fixed to a graded bronze meridian that swivels in all directions. The elegant globe features detailed cartography on paper map gores. It presents a dual map, alternatively presenting a political or physical map as the light is switched on or off.

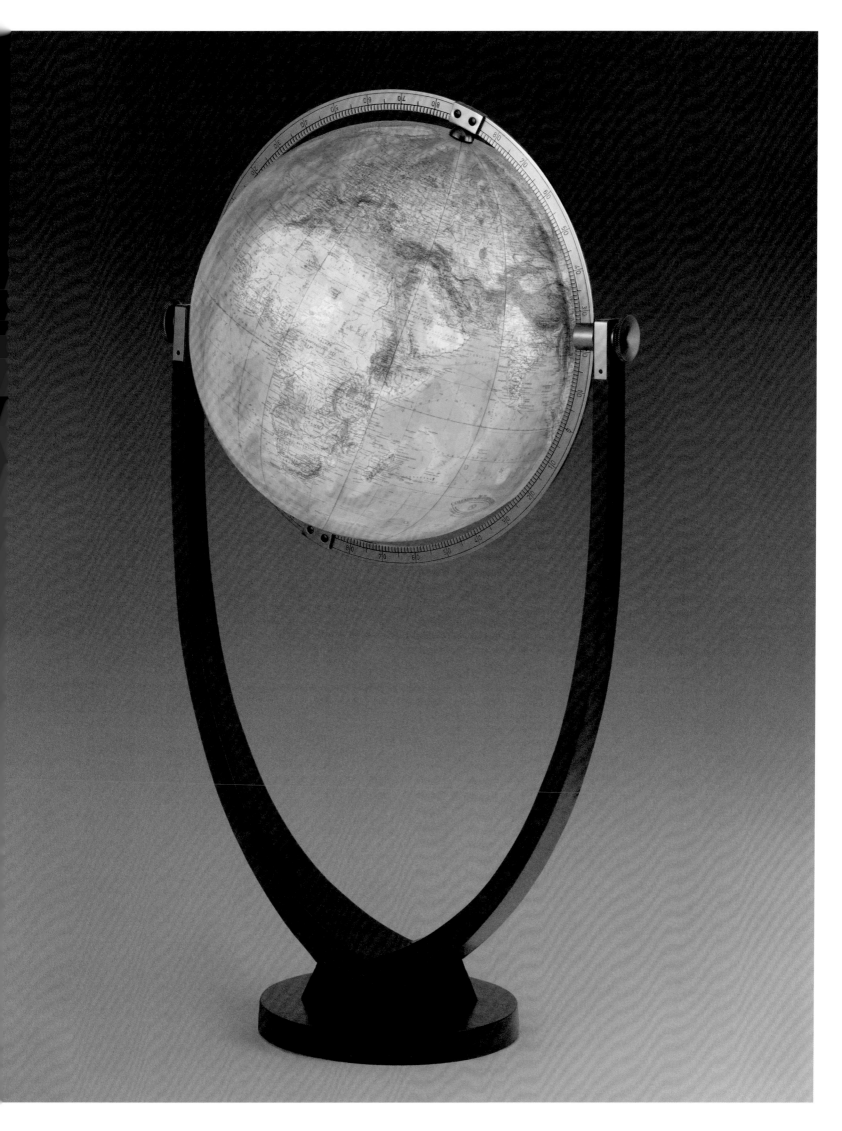

PAUL RÄTH

Wall Street Crash

Paul Räth (1881-1929) founded the Paul Räth Buchhandlung und Lehrmittel-Werkstätten in Leipzig in 1917. He was an entrepreneur who sold hundreds of educational products. With the acquisition of Otto Börner's globe factory, he became one of Germany's important globe makers. Räth's globes came in twenty languages, ranging from school globes to luxury globes. Räth made important contributions to globe innovation, such as the illuminated globe and the relief globe [no. 51]. In 1929, the stock market collapsed, fueling the onset of the Great Depression. Räth went bankrupt and committed suicide. The German population, burdened by hyperinflation and war damage payments, experienced extreme poverty. This in turn would create a breeding ground for the Nazi Party and would ultimately culminate in the Second World War.

Bombs and Communists

Managing director Rudolf Zschörnig restarted the company after Räth's death, but its return would be short-lived. During the war, the Allies bombed the Leipzig Book District. More than fifty million books and two thousand publishing and printing companies went up in flames. In the end, Leipzig was occupied by the Soviets, and was incorporated into communist East Germany. Zschörnig published a number of globes under the name Erzett, after his initials. The company was nationalized, its activities were restricted to globe production, and its name was changed to VEB Räthgloben Verlag, while VEB Hermann Haack printed its map gores. Räthgloben flourished during the GDR period, with eighty employees and a six-million-mark turnover. The yearly output of one hundred thousand globes was exported to fifteen different countries, including as far as Vietnam [nos. 148, 156].

Fall of the Berlin Wall

Räthgloben kept pace with time. In 1960, the company published a remarkable Moon globe, based on the Soviet lunar explorations [no. 101]. Inspired by the Trabant, the iconic East German car with a full plastic body, it started production of full plastic globes [no. 140]. In 1989, when the Berlin Wall came down, the East European market imploded. Räthgloben struggled on the global market, and its production dropped to twenty thousand units. American globe maker Cram bought Räthgloben to get a foothold in Europe, but the situation deteriorated, and the company went bankrupt again in 1999. Columbus Verlag hired some its staff, while Italian globe makers Nova Rico and Tecnodidattica acquired the publishing rights. They founded Räthgloben 1917 Verlag GmbH and incorporated it into their group, using the Räthgloben brand for a specific line of globes.

Räths Globes
Advertisement
/ Paul Räth
Lehrmittel-
Werkstätten /
Leipzig, 1923

140

Räthgloben, based in Leipzig, was the most important globe maker in the GDR. The company kept pace with industrial innovation and started producing plastic globes. This model from the 1980s is a full-plastic model. The eighteen-centimeter sphere is made of vacuum-molded plastic, with cartography printed directly on it. The meridian and the base are also cast in plastic. The brownish and greenish elements are sober and functional, a good example of East German design, a recognizable style in its own right, currently experiencing revived interest due to a recent wave of Ostalgia.

HERMANN HAACK

Physisich-Politische Einheitsglobus

Justus Perthes (1749-1816) established a publishing house in Gotha, Germany, in the late 1800s. For generations, Perthes remained the premier German map maker, publisher of the German Standard Atlas. Its globe production started in the twentieth century, when the young cartographer Hermann Haack (1872-1966) joined Perthes. Haack was dissatisfied with the poor cartographic quality of German school globes and endeavored to develop a superior alternative. With unparalleled precision, Haack drew map gores, incorporating thirteen thousand topographical names and thirteen color shades for height levels. He divided the information across two globes: a physical and a political globe. The pair, published in the 1910s, were a commercial failure: school budgets only allowed for the purchase of a single globe. Haack had to consolidate the information back into a single globe, the Physical-Political Unified Globe.

Nationalization

Since Perthes produced maps for the German Air Force, it was declared a Kriegswichtig Betrieb by the Nazi regime. Joachim Perthes (1889-1954), who was not a party member, had to accept a Nazi loyalist as manager to oversee him. At the end of the war, the U.S. Army occupied Gotha and proposed the family to move to the West with their maps, but they declined. Three months later, Gotha was integrated into the Soviet zone. With minimal war damage, Perthes quickly resumed production, yet in 1952, the East German government nationalized the company. Without compensation, the family established a new publishing house in Darmstadt. The communist authorities appointed retired Hermann Haack to manage the company in Gotha. As a corresponding member of the U.S.S.R. Geographic Association, he was considered a reliable person.

Privatization

The new company was renamed VEB Hermann Haack Geographisch-Kartographische Anstalt Gotha. It held the monopoly on cartographic school publications in East Germany, including printing map gores for various models of globe Its cartographic precision was complemented by the production skills of VEB Räthgloben, which manufactured the globes. The company flourished: during the 1980s, it employed a staff of 250 people with a turnover of eighteen million mark. In 1989, the Berlin Wall fell, Germany was reunited and the Perthes family was able to reclaim their former company. In 1992, it merged with the Darmstadt branch as Justus Perthes Verlag Gotha GmbH. That same year, the family sold the company to Klett Verlag Group. When the company was moved to Leipzig in 2016, 230 years of geographical tradition in Gotha came to an end.

VEB HERMANN HAACK GEOGRAPHISCH-KARTOGRAPHISCHE ANSTALT GOTHA

Craft a Globe
Cover page of a cut-out cardboard globe / VEB Hermann Haack / *Gotha, 1980s*

141
In the 1950s, VEB Hermann Haack Geographisch-Kartographische Anstalt in Gotha, Germany, designed this notable model, a monumental floor globe on a wooden cradle stand. It resembles an anthropomorphic sculpture, reminiscent of the abstract sculptures of Alexander Calder (1898-1976). This edition is remarkable, as wooden cradle stands were not commonly used by European globe makers. The sixty-four-centimeter sphere features physical cartography with pale-green oceans and yellowish landmasses, harmoniously complementing the warm tone of the wooden stand.

"Cartographers don't lie, they take a position. Maps aren't faithful portraits of reality, but subjective constructions. All maps are good, but they are all different, and in this difference you get a glimpse of our past and present."

Vladimiro Valerio, 2007
Expert in the History of Cartography
Venice University

Indonesian President Suharto (1921-2008), in traditional dress, shows a globe with south on top, a political statement towards its former North European colonists / 1967

DALMAU

Education and Games

Josep Dalmáu Carles (1857-1928) was a passionate teacher from Girona, Catalonia, an innovator in education, and even a builder of his own school. He founded a publishing house in 1904 and partnered fifteen years later with his son-in-law Joaquim Pla i Cargol (1882-1978). The family business, Dalmau Carles, Pla, S.A., became an important Spanish publisher of encyclopedias and educational books during the twentieth century. In the 1920s, Dalmau started its own globe production in collaboration with Sogeresa, a company specializing in didactic material. Dalmau produced various sizes and models, such as art deco-styled globes. The color-lithographed paper map gores displayed refined cartography, with appealing colors and an elegant wavy pattern to visualize the ocean currents. A curved metal pin fixed the spheres to a large wooden disc-shaped base [no. 142].

Chrome and Varnish

After the war, Dalmau transformed its globes to a modernist style. It updated the cartography with the latest postwar political borders and enhanced the map's colors with a fresh turquoise for the oceans and yellow ochers for the landmasses. A thicker varnish layer was applied to achieve a stronger shine. The spheres were mounted on a chromed steel wire tripod, comparable to the French globes of the same period. In the 1970s, the Spanish market saw the rise of German-made illuminated plastic globes. To catch up, Dalmau soon began its own plastic globe production. During the 1980s, Dalmau sold various plastic globes with mythical names such as Lume, Relieve, Nadir, Discovery, Orion, Zenit, and Orbis. With a production of ninety thousand units and a turnover of ninety million pesetas, Dalmau's globe department was a successful branch of the company.

Eight-Hour Day

Grandson Joaquim Pla i Dalmau (1917-2005) took over the company in the 1960s. One of the social achievements after the war was the 8 8 8 principle: eight hours of work, eight hours of relaxation, and eight hours of sleep. Dalmau understood the Zeitgeist: once acquired, relaxation time had to be filled. He shifted to board games but continued the family's educational mission by linking education to amusement. When television conquered Spanish society, Dalmau responded by designing educational board games based on Spanish television programs, such as the popular TV quiz *Un, dos, tres, responda otra vez*. Dalmau also created Spanish versions of popular American computer games, such as *Reconquest*, or *Dungeons & Dragons*. Globe production ceased when Dalmau sold the family business to printing company Alzamora Artegráfica SA in 1981.

The World of Things
Cover of an educational book / Published by Dalmau Carles, Pla, S.A. / Girona, Spain, 1960s

142
Dalmau, the globe maker from Catalonian town Girona crafted this striking globe in the postwar period. The thirty-five-centimeter sphere can be freely observed from all sides, as it has no horizon ring or meridian. It is fixed only to a bent metal pin, which slowly rises out of an enormous circular wooden base—achieving a delicate balance between fragility and stability. Its Spanish-language map features a vibrant wavy pattern that aptly visualizes the ocean currents. The greenish oceans and yellow-ocher countries create a nicely balanced color scheme.

VALLARDI

Tourists and Railroads

Francesco Cesare Vallardi (1736-1799) continued his uncle's bookstore, a family business that would last for six generations. In eighteenth-century Milan, the shop was an artistic hangout frequented by writers from the so-called Lombard Enlightenment. In the Romantic period, the tradition of the Grand Tour began, a coming-of-age trip to Italy for upper-class European young men. Vallardi responded to this new phenomenon by publishing the *Itinerario d'Italia*: the first tourist guide, featuring fold-out maps and tourist information, such as hostels, travel times, and the price of postal horses. Thanks to the new railroads, tourism surged during the nineteenth century, creating high demand for tourist guides. This gave Vallardi's publishing house a significant boost and led the company to open branches in London and Paris to be closer to their customers.

Tourists and Automobiles

Francesco Vallardi (1809-1895), of the third generation, was active in the Risorgimento, the movement for the unification of Italy. A doctor by profession, he founded a publishing house that specialized in medical, cultural, historical, and encyclopedic works. His brother Antonio Vallardi (1813-1876) started Antonio Vallardi Editore, focused on geography and education. It specialized in teaching materials such as atlases, schoolbooks, wall charts, and globes. When the automobile appeared on the streets of Italy around 1900, the company successfully launched road maps and travel guides in collaboration with the Touring Club d'Italia. Fifth-generation Antonio Vallardi (1882-1965) consolidated the company's dominant position in the Italian market in the twentieth century and initiated international expansion by selling publications in Europe, Africa, and Latin America.

Bakelite Fins

Antonio Vallardi Editore made globes from the 1920s to the 1960s in many shapes and forms, ranging from ten to forty-five-centimeters in diameter, with cartography drawn by A. Minelli. A relief globe was also included in the assortment. Vallardi's small school globes stand out for their sober design. They lack a meridian and are mounted directly on the South Pole. The oldest models have a symmetrical wooden tripod stand [no. 59], the later ones an asymmetrical wooden leg [no. 60], and eventually a curved Bakelite fin, creating a very dynamic image [no. 143]. In 1970, Garzanti Publishers took over the Vallardi family business but sold back the cartographic rights to Giuseppe Vallardi (1925-1993), the sixth generation, who founded Vallardi Industrie Grafiche SPA. The company continues the family tradition of making maps and atlases, but no longer produces globes.

Didactical Materials
Cover of Physics Catalog n.9d / Antonio Vallardi Editore / *Milan, 1950s*

143
Vallardi, the Italian globe maker from Milan, created this model in the 1940s. It was offered both as a terrestrial globe and as a celestial globe, featuring a star map by astronomer Prof. E. Sergent (x-1895), with constellations depicted as mythological figures on a pale-green background. The eye-catching feature is the streamlined art deco stand, cast in one piece, an early example of a monolithic Bakelite stand. The curved fin creates a dynamic image, rising in a single movement from the round base towards the inclination of the Earth, balancing the twenty-centimeter sphere on its tip like a ball on the nose of a seal.

PARAVIA

Fighting Illiteracy

Giovanni Battista Paravia (1765-1826) founded G.B. Paravia & Co in Turin 1802, a publishing house specializing in religious and scholastic textbooks. His son Giorgio (1796-1850) and nephew Innocenzo Vigliardi (1822-1896) continued the business at a time when most of the European population was still illiterate. Governments imposed various laws to improve education, such as the obligation to purchase teaching materials like books, wall charts, and globes. Paravia focused on the booming school market and saw steady growth. Carlo Vigliardi Paravia (1845-1919), the eldest son of Innocenzo, was sent to France to study the progressive education system before heading the family business. The unification of Italy opened new markets, leading Paravia to establish new branches in Rome, Florence, Milan, and Naples, each one led by one of Carlo's brothers.

Colonialism and Fascism

At the Colonial Conference in Berlin in 1884, the European powers divided Africa among themselves without inviting a single African representative. Italy appropriated Libya, Eritrea, and Somalia. In the 1920s, the fascist government of Benito Mussolini (1883-1945) set up a program to promote public enthusiasm for colonialism. It included wall maps to teach schoolchildren about the glorious Italian colonies and fascist themes. For commercial reasons, Paravia aligned itself with the regime. It produced a number of school charts with roaring names such as 'The Expression of the Struggle for Integrity of the Race', drawn by Roberto Almagià (1884-1962), who ironically ended up on the list of unwanted authors following the racial laws of 1938 because he was Jewish. After Italy lost the Second World War, its colonies were seized by the victors and no longer shown on globes.

Italian Rationalism

Paravia began publishing globes in the early twentieth century, starting with traditional models featuring wooden stands and paper and plaster spheres. Cartographer Guido Cora (1851-1917) designed the map gores. In the 1930s, Paravia's globes evolved into an elegant rationalist style, with a metal base and a gimbal suspension. Tancredi Vigliardi Paravia (1884-1969) and his son Carlo (1914-2006) rebuilt the company from the ruins after the Allied bombings. They restyled the globes, giving them rich colors and varnishes, and mounted them on modernist metal tripods, suited to mid-century Italian interiors. The cartography was refreshed and updated. Paravia became one of the most successful Italian publishers, employing a staff of two hundred. In 2000, the company merged with Milanese publisher Bruno Mondadori, and was bought by the British Pearson Group six years later.

Obligatory and Recommended School Materials
Cover of Paravia Catalog /
Turin, 1924

144
Paravia in Turin made a special globe model called Tellux. The name refers to a Tellurium, a mechanical device that imitates the movements of the solar system. In this case, it consists of a thirty-centimeter terrestrial sphere accompanied by a small wooden ball representing the Moon, mounted on a steel wire. Its position can be shifted to demonstrate the relation between Sun, Moon, and Earth in the various seasons. It is patented by C. Boehmer, who designed more globes for Paravia. This modernized version from the 1960s has a heavy metal disc as a base and was also available with a metal tripod stand.

RICO &
TECNODIDATTICA
GLOBES

Plastic Pioneers

Rico was founded in 1955 in Impruneta, close to Florence, Italy. The company originally produced maps and office stationery, but from the 1970s onward, it focused entirely on globes. As a young company unburdened by tradition, Rico pioneered the industrial production of plastic globes. Its two hemispheres are machine-made by a vacuum press and wrapped in a thin transparent foil with a map printed on it. This is heated to sixty-five degrees and modeled onto the sphere. Finally, the globe is placed in a plastic meridian and fixed to a plastic base. Over time, Rico made various models, including political and physical globes, relief and illuminated globes, and even Moon globes, printed in many foreign languages for export. Rico's globes stand out for their unpretentious aesthetics, radiating the optimistic era of the 1960s [no. 9].

Shop to Factory

Riccardo Donati opened an office supply store on the Via Stendhal in Milan and started Tecnodidattica in 1949. Together with his son Renato Donati, he manufactured office stationery and globes. In the early years, they sought new techniques, materials, and designs to develop a contemporary globe. The globes were made industrially, entirely from plastic using the vacuum press process. Riccardo loved the sea more than the city, so he moved the business to the quiet Ligurian Coast, where he build a new factory near San Colombano Certenoli in the 1970s. Tecnodidattica expanded to become an important globe maker. The 1990s were a turbulent period for globe makers. Many takeovers, mergers, and bankruptcies led to a reshuffling of the industry and the globalization of the globe business, of which Tecnodidattica would emerge as one of the winners.

Mergers and Takeovers

Rico and Tecnodidattica merged in 1996 and embarked on an expansionist strategy under the direction of grandson Riccardo Donati and Stefano Strata. They acquired other European globe makers, such as the German Räthgloben and the Danish Atmosphere. The company also obtained the European license to produce globes for National Geographic. Today, the group has become one of the largest globe producers, with an annual production of eight hundred and fifty thousand globes, eighty percent of which are destined for export. The company has a distribution network in seventy different countries and produces globes ranging from school models to luxury objects, from eleven to one-hundred-twenty-eight centimeters, in forty different languages. The brand names Nova Rico, Tecnodidattica, Räthgloben and Atmosphere are used for different lines of globes [no. 189].

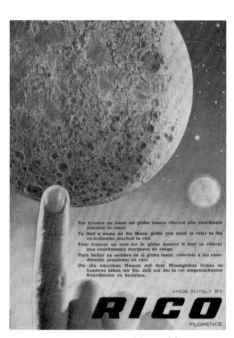

**Manual for
an Illuminated
Moon Globe**
Rico / Florence,
1960s

145
In the 1950s, Rico pioneered the industrial production of machine-embossed plastic relief globes in the Florence region of Italy. A negative metal mold is created with the height differences milled inside.
A flat acrylic sheet is color-printed with the cartographic image, pre-heated and molded under pressure into a relief hemisphere. Two hemispheres are then glued together to form a plastic relief sphere. In this case, Rico's relief globe features a thirty-centimeter sphere with a Farbenplastik map coloring, which enhances the relief effect. The stand consists of a graded plastic meridian and a steel wire tripod.

SCAN-GLOBE

From Kiosk to Multinational

Danish Alfred G. Hassing (1890-1939) started a bookstore in Copenhagen in 1915. Initially, it was just a kiosk in the lobby of the *Dagbladet Politiken*, a newspaper office on Rådhuspladsen. Hassing soon expanded his business into one of the largest bookstores in Scandinavia, called Boghallen. He founded a publishing company called Hassing Forlag that also sold globes, initially called Boghallen Globus and later known as Tower Globus. The small workshop produced traditional models mounted on a central wooden stand, with an output limited to only four globes per day. The postwar director of Hassing, Willy Schmidt (1913-2002), achieved a breakthrough in 1963: a joint-venture with Replogle from Chicago, the world's biggest globe producer at the time. The new company, called Scan-Globe A/S, was relocated to Havdrup, a small town near Copenhagen.

Atlantic Joint Venture

Globe markets in Europe and America have historically been separate, each with their own dynamics and models, apart from some British-American exchanges of map gores at the beginning of the century. As such, the intercontinental joint venture presented an opportunity to both sides. The Danes, known for their famous mid-century modern design, brought innovation and creativity, while the Americans brought their experience in mass production and marketing. Hassing could professionalize its production, while Replogle gained a foothold in Europe. By the 1980s, when Per Lund Hansen (1939-) succeeded Schmidt as director, the company had become the largest European globe factory, producing popular globes with an annual output of six hundred thousand units. Models were available in twenty-four languages, essential in Europe's multilingual market.

Shareholders and Bankruptcy

Scan-Globe designed many innovative models. It patented the Spot-Globe, which could spotlight any location by tuning knobs to a specific latitude and longitude [no. 29]. It pioneered Planet Earth, the first globe using a satellite photo as a map [no. 186]. It also created stylish curved Plexiglas stands and produced Moon, Mars, and celestial globes [nos. 43, 45, 103, 105]. In 1988, the Danish partners sold their shares, leaving the Americans as the sole owners. A year later, the Berlin Wall fell, borders began to shift, and globe sales plummeted. Scan-Globe faced difficult years, surviving by selling cheap plastic globes. In 2003, the American owners filed for bankruptcy, dismissed the fifty-five Danish employees and shipped the machines to Chicago. Per Lund Hansen started a new globe company called Atmosphere, which was purchased six years later by the Italian company Tecnodidattica.

Globus 2000, The Illuminated Globe with the "Jumping Point"
Globe Manual
Scan-Globe /
Copenhagen, 1970s

146
Scan-Globe, a joint venture between Hassings Forlag in Copenhagen and Replogle from Chicago, produced innovative globes in the 1970s. This particular model has a thirty-centimeter illuminated sphere with paper map gores, displaying a French-language physical map. The cartography includes the ocean floor's relief. The sphere rests on a very unusual base, crafted from a block of polished teak wood. The sculpted stand, impossible to manufacture as a mass-product, is likely a custom-made piece. It gives the globe a unique expression, making it a great conversation piece in any living room.

More European Globes

147

Bolis, the Italian publishing house from Bergamo, created this elegant, illuminated globe on a marble base as an object for interior decoration in the 1950s. The fourteen-centimeter glass sphere appears to be delicately supported by a single gold-colored tube. It achieves a precious equilibrium, a clever effect that enhances the idea of Earth floating in space.

148

Räthgloben, the nationalized East German globe maker in Leipzig, produced many globes. This 1950s model had a broad distribution. It combines a twenty-one-centimeter sphere with paper gores displaying postwar cartography and a Bakelite stand. A cast-metal airplane is mounted on top, likely a later addition and not part of the original design.

149

Columbus Verlag in Stuttgart manufactured quality globes in the postwar era, such as this elegant 1950s table model in a decorative modernist style. The Duo-Leucht Globus, a thirty-six-centimeter double-image illuminated glass sphere, rotates effortlessly on a curvy brass tripod stand. The cartography is very detailed, featuring a wealth of information.

150

Road map maker JRO in München also produced luxury globes after the Second World War. This model, designed in the classic modern style of the 1950s, has a twenty-five-centimeter sphere of paper gores featuring densely informative cartographic details. It is mounted on a metal meridian, which sits on top of an elegant wave-shaped Bakelite base.

151

Dietrich Reimer, a Berlin globe maker, published this globe under the name Berolina in the 1950s. Its thirty-centimeter sphere with paper gores shows Africa divided by the colonial powers, a few years before the start of decolonization. The balanced pastel map colors, shiny meridian, and dark-wooden base make it an elegant object.

152

Columbus Verlag made this streamlined globe in Stuttgart in the 1950s. It is a Duo-Leucht Globus, a thirty-six-centimeter double-image illuminated glass sphere. It rests on a heavy metal base, featuring a single leg that combines the meridian and base in one continuous movement. The political-physical map has refined graphics.

More European Globes

153

Map and globe maker Taride in Paris commissioned the well-known designer couple Audoux & Minet to create a special globe in the 1950s. The pair used a standard fifteen-centimeter illuminated glass sphere from Taride with paper gores and placed it on a stand made of abaca rope, in the same style as their laidback modernist furniture on the Côte d'Azur.

154

This terrestrial metal globe was produced by an unknown Italian company in the 1960s. The bright cartography is lithographed on the twenty-four-centimeter sphere. This diameter is remarkably large for a tin globe. The sphere rests on a single steel wire, which is bent into the shape of a triangular cantilevered stand.

155

This Italian globe was made by Laboratorio Italiano Mappamondi in Torino in the 1960s. The map gores on the thirty-centimeter sphere display cartography with a physical-politicalmap. The white color indicating sea depths produces a confusing, fragmented ocean image. The sphere is mounted on a graded meridian and a metal tripod stand.

156

The East German Cartographic Institute VEB Hermann Haack in Gotha printed the map gores for this twelve-centimeterglobe, made by Räthgloben in Leipzig in the 1960s. The model was customized as an advertisement globe for the East German Interflug airline, by painting the wooden base in the company's red color and glueing its logo onto it.

157

This brightly colored globe was made by Taride in the 1950s. The twenty-five-centimeter cardboard sphere with paper map gores is mounted on a metal meridian that rests on a tripod of whimsical steel wire legs. The legs end in playful white rubber ball feet. It is a typical joyful object for an optimistic and carefree period in France.

158

Michael Seidel from Zirndorf in Germany started as a tin toy factory but gradually transformed to a producer of plastic toys. This 1960s globe is a hybrid: the twenty-centimeter illuminated sphere is made of vacuum-molded plastic with a geographical print, the graded meridian is of transparent plastic, while the red base is still made of metal.

AMER-
ICA
&U.K.

World Domination

At the onset of the twentieth century, the United States adopted isolationist policies. However, when the Japanese attacked Pearl Harbor in 1941, the U.S. entered the Second World War and ultimately emerged as one of its victors. It marked the beginning of a long period of American global dominance, matched only by that of the Soviet Union.

During the presidency of John F. Kennedy (1917-1963), the world faced perilous geopolitical crises, such as the conflict over the deployment of Soviet missiles in Cuba. To analyze the geopolitical situation at hand, a monumental presidential globe always graced the Oval Office of the White House. Globes were often exchanged as gifts among statesmen; Kennedy presented a 32-inch Replogle globe to the newly independent African state of Tanganyika on its Independence Day in 1961. Tragically, two years later, Kennedy was assassinated in Dallas during an open limousine tour.

The image depicts Kennedy in the Oval Office making a telephone call next to his large, illuminated globe. This gift, from Admiral Arleigh Burke (1901-1996), Chief of Naval Operations, is a 24-inch Replogle model known as the Executive Classic Globe. Mounted on a metal meridian and fixed to a wooden base resembling a side table, this globe is a departure from the ornately decorated presidential globes of the past, signaling a new era of sobriety and modernity. It was a delightful object for the president's young children to play with when they invaded their father's office.

While British globe makers continued to build on a long tradition, American globe makers started in the early twentieth century as individuals pursuing the American dream. Each of them founded their own company with its own history. They sold globes out of the trunk of their car or set up a small workshop in their in-laws' garage. The American entrepreneurs built a global industry from the ground up while navigating through economic crises and world wars.

PHILIP & SON

Titanic

Scotsman George Philip (1800-1882) started his career as a book-seller in Liverpool. In response to growing demand for educational materials, he expanded into teaching materials, maps, and atlases. The company relocated to the Caxton Building, a huge complex with motor-driven presses for printing lithographic-colored books, maps, and even its own paper. At its peak, Philip employed over three hundred people. His son George Philip Jr. (1823-1902) guided the company into the twentieth century. In those days, ocean liners such as the Titanic, the largest ship in the world, were the only means to travel between Europe and America. The Titanic was registered in Liverpool, the primary Atlantic port handling forty percent of world trade. Despite Philip & Son's impeccable sea charts on board the Titanic, the ship struck an iceberg and sank on its maiden voyage in 1912.

Georama

Philip & Son commenced globe production around 1900, with assistance from the London Geographical Institute. To gain expertise, it acquired traditional British globe makers like Malby, Betts, and W. & A.K. Johnston. Consequently, Philip's early globes look outdated. In the 1930s, the company modernized the cartography and started using art deco-styled metal and Bakelite stands [nos. 37, 159]. Many globes were destined for export, as expressed by this ad: *"The Challenge Globe, an Overseas Model in eight languages. Its reception proved that more than ever before, interest in world affairs has made the possession of a globe a necessity."* The acquisition of globe maker Georama provided Philip & Son with some innovative models, including an inflatable vinyl globe, an illuminated plastic sphere, and a cradle globe [no. 180]. Unfortunately, though, these advancements came too late.

Cravats and Scissors

In 1955, a British Pathé cinema newsreel showcased the production of globes at Philip & Son. A craftsman wearing a cravat is seen meticulously smoothing nine layers of plaster on a papier-mâché sphere and passing it to a well-groomed lady. She painstakingly scissor cuts paper map gores, glues them onto the sphere, and delicately applies retouches with a fine brush. Each globe required fifteen hours to complete. While it was the intention to highlight the pride of British globe making, it signaled Philip & Son's impending demise. In the meantime, on the other side of the Channel, globe makers invested in machines to mass produce plastic globes at a rate of one per minute. In 1987, Reed International acquired Philip & Son and ceased globe production. The cartography department was split off and Georama regained its independence. Reflecting the trend of globalization, French publisher Hachette acquired British Philip & Son in 2001.

Cover page of Elger's Map of the Moon
George Philip & Son / *London, 1959*

159

George Philip & Son, the British globe maker, crafted traditional globes, including this refined celestial model. It features a six-inch Bakelite sphere with dark-green paper map gores. It shows the night sky with golden stars of magnitudes one to five. The constellations are depicted as interconnected line figures. The stylish model was made from the 1920s to the 1950s, available in different sizes and colors, and even as a set paired with a terrestrial globe. Initially, the celestial sphere was mounted on a traditional brass stand, later transitioning to a stepped art deco-styled base crafted from black Bakelite.

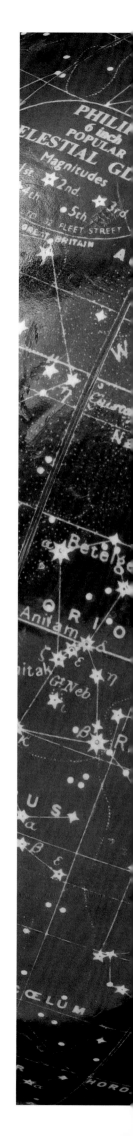

WIGHTMAN

Hobby sculptor

British sculptor Arthur J. Wightman (1906-1977), a retired electrical engineer, shared a passion for sculpting with his wife, Gertrude. Together, they spent two years sculpting a relief globe of the Moon. Utilizing chicken wire, cotton rags, and fiberglass as the basis, they applied a top coating of filler and emulsion paint. Wightman carved the lunar relief, drawing inspiration from photographs captured by the American Lunar Orbiters, which he obtained from NASA. Wightman drew a negative rubber mold of the master globe, enabling him to cast copies by pouring plastic into the mold. His handcrafted so-called Lunaglobe looked remarkably realistic, featuring many craters and plains. In a bold publicity move, Wightman journeyed to Buckingham Palace in London the day after Apollo 11 had landed on the Moon, to present a copy of his Lunaglobe to young Prince Charles (1948-), newly crowned as heir to the throne.

The Man Who Sold the Moon

The public appetite for lunar globes surged in July 1969, overwhelming Wightman's capacity to keep up. Lunasphere Productions Ltd., his one-man company, had an output of a mere ten globes a week. A British Movietone newsreel titled *The Man Who Sold the Moon*, shows Wightman in his workshop in Penzance, Cornwall, hand pouring and hand painting Moon globes. In an interview, he proudly exclaims: *"The Americans are making Moon globes, but they admit ours is the Rolls-Royce of Moons!"* He boasted about receiving orders for a million globes from the U.S., about setting up a factory, about hiring thirty collaborators and producing twenty thousand globes a week. Wightman's lunar globes made appearances in the James Bond movie *Diamonds are Forever* and adorned the homes of Apollo 8 astronauts Frank Borman (1928-2023) and Jim Lovell (1928-).

Do-It-Yourself Moon

Wightman's Lunar Globe came in three diameters: one, two or four feet, mounted on a cross-shaped plastic cradle base, with the company's name molded into it. The standard hand-painted globe was priced at $12, while the larger globes, intended for museums and universities, were priced at $120. To boost production, Wightman sold blank spheres at $4.90, which customers had to paint and finish themselves. Rub-off sheets with the names of craters and plains were provided for customers to apply, along with red and yellow dots to mark the landing sites of Soviet and American spacecraft. As a result, every Wightman globe is unique, hand finished by its first owner. It remains unclear what became of Wightman and Lunasphere Productions. His globes are now rare finds, so it is doubtful that he ever delivered the promised one million copies.

Packaging box for Wightman's Lunaglobe
With Key Chart & Index / Wightman Lunasphere Productions Ltd. / Penzance U.K., 1969

160
Arthur J. Wightman, a British amateur globe maker in Cornwall, crafted lunar relief globes from cast resin. This model features a twelve-inch sphere, with lunar plains depicted in dark grey, the basic color of the plastic material used for the globe. The yellowish mountain ranges and craters are meticulously hand painted. Craters and plain names are applied using rub-on letters, while red and yellow paper dot stickers denote the landing sites of the Soviet and American spacecraft. The globe is supported by a plastic cradle base, shaped as a slide-in cross, bearing Wightman's brand name cast into it.

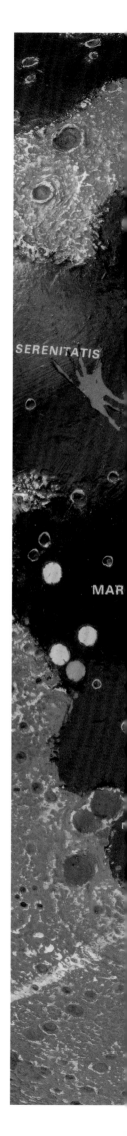

HAMMOND

Bibles and Almanacs

Caleb Stillson Hammond (1862-1929) managed the New York office of Rand McNally. However, one day in 1900, when he was refused a five-dollar raise, he walked out and set up C.S. Hammond & Co. His family business would endure for four generations, producing cartographic items such as maps for encyclopedias, almanacs and bibles, along with atlases and globes. Initially, Hammond sourced map gores from Scottish map maker W. & A.K. Johnston and glued its label onto the cartouche, a common practice due to the complexity and cost of creating map gores. By the 1920s, C.S. Hammond & Co. began crafting remarkable globes based on their own map gores, often mounted on elegant art nouveau cast-iron stands [no. 24]. Other models of that time had a wooden base containing a Hammond Atlas, a clever way to sell two products at the same time.

Terrascapes

In the 1930s, Hammond introduced several art deco-styled globe models. These featured detailed map gores, a sober Bakelite base, a metal meridian, and a glossy finish, resulting in globes that were visually pleasing with a warm feel [no. 104]. After the Second World War, grandsons Caleb Jr. (1916-2006) and Stuart (1922-2015) relocated the family business to Maplewood, New Jersey. Hammond became the second-largest map maker in the U.S., with a yearly output of one billion maps a year and a staff of more than one hundred. In the 1960s, chief cartographer Ernst Hofmann (1922-) developed a shadow relief globe. He created so-called Terrascapes, geographical scale models made of thin aluminum plates, photographed them and transferred the photographic shadow onto the map to create the illusion of height. Hammond also produced an eighteen-inch plastic relief globe.

Computer Cartography

Hammond employed ten researchers and many correspondents worldwide to collect geographical data to keep their maps up to date. On one occasion, they had to wait six months to receive data from India because their contact had been on an elephant hunt. In the 1980s, great-grandson Caleb Dean Hammond III (1947-) spearheaded computer-generated cartography. Investing in a one-million-dollar computer system, he developed a geographical database and produced the first digital atlas. Hammond's digital maps gained recognition when they were used by CNN during the live broadcast of the U.S. bombing of Bagdad in 1991. Soon, digital cartography would become the map industry's standard. At the turn of the millennium, the family sold its business to the German Langenscheidt Publishers, which rebranded it as Hammond World Atlas Corporation.

Cover of Hammond's Self-Revising World Atlas and Gazetteer
C.S. Hammond & Co. / New Jersey, 1940s

161

C.S. Hammond & Co. in New Jersey marketed this sixteen-inch inflatable globe in the 1950s as a floor model and a table model. The vinyl map gores featured detailed cartography, and valves are used to fix it to the poles and enable it to rotate on its lightweight steel wire stand. Marketed as a serious educational tool, it was sold for $9.95 and offered for $3.50 at Amoco gas stations: *"Here is a valuable educational aid your family will refer to again, and again. A decorative piece of charm and interest. A timely back-to-school purchase. A fine gift to set aside for Christmas or birthdays".*

NYSTROM

Assembled in Five Minutes

Alfred J. Nystrom (1873-1956), of Swedish descent, established a school supply company in Chicago in 1903. Initially, he used map gores from the British W. & A.K. Johnston company, featuring complex cartography for adults. However, Nystrom pioneered the idea of simple globes tailored to the cognitive level of children, affordable and robust, to withstand the rigors of the school environment. From the 1910s, Nystrom produced six-inch wire stand globes, designed to sit on every child's desk. In the 1920s, Nystrom patented the Form-a-Globe, a twenty-five-cent package of pre-printed paper map gores that could be *"Assembled in Five Minutes"* [no. 193]. Another educational idea was the Suspension Globe, suspended from the ceiling of a classroom and lowered for demonstrations by means of pulleys and ropes, including a counterweight disguised as the Moon.

The Hitler Headache

Nystrom's educational catalog also offered Black Slate Globes and Project Globes, allowing students to chalk maps for projects and training purposes [no. 48]. Nystrom's globes can be recognized by a patented metal disc at the North Pole, a time dial that indicates time zones. Alfred Nystrom retired shortly before the Second World War, a difficult time for globe makers. *The Chicago Tribune* ran an article titled *The Hitler Headache*, in which his successor C.A. Burkhart was interviewed about the necessity to revise globes following the annexation of Austria and Sudetenland by Nazi Germany: *"... he is hoping things will be settled so that his company may begin production of new terrestrial globes. Eventually the situation will help business, but that will take about two years, he's afraid"*. In reality, Nystrom had to wait seven years before it could revise its globes to reflect postwar borders.

Computed Maps

"Many new Babies Assure Globe-maker's Future" was a headline in the 1950s. Nystrom prospered after the war, selling school globes with basic maps and large letters, such as the Beginners Simplified Globe. The third director, Charles B. Stateler (1901-1987), gained expertise in relief globes by acquiring the Raised Relief Map Division of the Aero Service Corporation in the 1960s. The result was the Pictural Relief Globe, which used shadow effects for height illusion, or the Raised Relief Globe which utilized relief cast in plastic [no. 21]. In the 1980s, Nystrom purchased its first Apple computer, with a graphic component to make the transition to digital maps, drawing them in half the time, with half the labor. In the 1970s, Herff Jones bought Nystrom, which later became part of Social Studies School Service, a school supplier that still uses Nystrom as a brand name.

Maps - Globes - Charts
Cover of A.J. Nystrom & Co. Catalog / Chicago, 1920s

162

A.J. Nystrom & Co., the Chicago-based school supplier, specialized in educational maps and globes. This 1960s model features a sixteen-inch sphere with paper map gores, displaying a striking cartography.
The physical map employs stark contrasting colors and shadows to create a relief effect. Large and simple lettering ensures easy readability from a distance, ideal for classroom settings. While the globe exudes a somewhat traditional look with its heavy metal disc base, it is clearly designed as a robust and timeless teaching tool for use in demanding school environments.

DENOYER-GEPPERT

GLOBES

Professor meets Salesman

Levinus P. Denoyer (1875-1964) was a professor of geography at the University of Wisconsin. In 1913, globe salesman Otto E. Geppert (1890-1970) attended one of his lectures and persuaded him to join him in Chicago to work at map and globe maker A.J. Nystrom & Co. Three years later, they founded the Denoyer-Geppert Company; Levinus focused on designing maps while Otto took charge of sales. Initially, the company distributed globes from British manufacturer Philip & Son, but soon it developed its own highly successful Cartocraft Globe. After a few years, the company relocated to larger premises in the Swedish-American Telephone Building in Edgewater, Illinois. It promoted itself as producer of *"visual demonstration equipment for geography, history and biological sciences,"* offering not only maps and globes but also eight hundred different wall plates and anatomical models.

Hidden for the Eye

In a reflective mood in the 1940s, Otto E. Geppert wrote: *"... A map is the visual symbol of something that is too large or too hidden for the eye to encompass, ... England or Italy or Australia, don't you first think of its shape on the map? Yet nobody has ever seen that shape in its entirety...."* During this period, they successfully marketed the Aviation Globe, a large metal globe designed for military training. The company also created cradle globes for schools, featuring patriotic names such as Victory, Freedom, and Liberty, alluding to the desired outcome of the war [nos. 163, 178]. In the 1960s, Denoyer-Geppert joined the Space Race with the Vanguard Globe, which demonstrates a satellite's orbit by means of a wooden ring [no. 98]. The White House ordered lunar globes as gifts for foreign guests, recognizing them as the most accurate model available [no. 102]. In the 1970s, the company also produced a detailed Mars globe [no. 184].

Map Factory Lofts

Levinus P. Denoyer remained the company's president until his death in 1964, by which time sales had reached six million dollars. In Edgewater, there was a strong sense of community, as most inhabitants worked at the plant. Three years later, Otto E. Geppert died in a car crash, a common cause of death at the time with more than fifty thousand people dying in road accidents in the U.S. that year. Times Mirror initially acquired the company, followed by Rand McNally in the 1980s. They merged the cartographic departments, discontinued the globe brand, and spun off the anatomical model department as the Denoyer-Geppert Science Company, which remains active today. As the economy shifted from industry to services in the late twentieth century, many factory buildings became vacant. The Denoyer-Geppert plant was renovated into apartments and rebranded as Map Factory Lofts.

Geographic Knowledge for World Understanding
Cover of Catalog 22 / Denoyer-Geppert / *Chicago, 1947*

163

Denoyer-Geppert Company, a specialist in educational globes, created a number of cradle globes during the Second World War, which later became popular school globes. This model from the 1960s was named Victory. It features paper map gores affixed to a twelve-inch fiberboard sphere, supported by an asymmetrical wooden cradle base. The cartography is simplified for educational purposes, consisting of a physical colored map depicting natural features such as rivers and mountains. Overlaying this is a political map, with borders delineated in red lines and country names in large letters for easy reading.

"There are many wonderful, absolutely terrific things you can do once you get enough geographical information in a data base".

Conroy Erickson, 1988
Rand McNally Globes
Chicago globe maker

FARQUHAR

GLOBES

Bubbles and Canopies

Robert H. Farquhar (1902-1983), a production engineer, and his friend, horticulturist James Harlan Kugler (1903-1957), shared a passion for astronomy. During the Second World War, Farquhar stumbled upon a new invention: transparent Plexiglas bubble canopies for warplanes [see page 26]. This innovation inspired the two friends to solve a two-thousand-year-old problem: how to represent the universe on a globe. While traditional celestial globes depict the stars on their outer surface as if one observes the universe from outside, a transparent celestial globe allows one to observe the universe from under the dome, as seen from Earth. After the war, Farquhar resigned from his job, underwent six months of training in a Plexiglas factory, sold his house to raise funds, moved in with his mother-in-law and started a workshop in an old Philadelphia garage to pursue his dream.

Transparent Globes

By 1948, Farquhar's Celestial Navigation Sphere Device 1X3, was ready. Developed in collaboration with the U.S. Navy and Air Force, who ordered a few hundred copies for navigation training, the device featured stars printed on the Plexiglas sphere, with a small Earth and Moon ball hanging inside. Various parts could be manipulated to study the movements of celestial bodies [no. 96]. Ten years later, the company patented an improved version called Earth in Space [no. 164]. For four decades, the Farquhar Transparent Globe Company produced Plexiglas globes from one to four feet in diameter, both celestial and terrestrial [see page 134]. Coastlines, borders, and grade lines were all are serigraphed in vivid colors on the Plexiglas spheres. Students could color the globe with crayons and wipe it off afterwards, making it a popular teaching tool, purchased by many schools [no. 8].

Spherical Concepts

High-quality transparent globes for education remained the niche product of Farquhar's Transparent Globe Company. Operating as a small workshop, the company ceased to exist after the death of Robert H. Farquhar. General manager John M. Szal (1947-) and plant manager Mark A. Buikema (1954-) then founded Spherical Concepts Inc. in the 1980s, their own globe-making company. It continued the production of transparent globes similar to Farquhar's models, such as the Starship Earth Celestial Sphere or the Bowl of Night [no. 95], but also introduced a line of acrylic globes called Artline and more traditional globes for *National Geographic*. In 2006, Spherical Concepts was sold to Herff Jones, an educational products company that had earlier acquired globe makers Cram and Nystrom. Five years later, however, Spherical Concepts was dissolved.

The Farquhar
Transparent
'Tool' Globe
Promotion
Postcard /
Farquhar
Transparent
Globe Company /
Philadelphia, 1950s

164
The Farquhar Transparent Globe Company in Philadelphia introduced this radical model in the 1950s, known as the Earth in Space™ Celestial Globe. Featuring a twelve-inch astro globe and a double curved base, both crafted from transparent Plexiglas, this globe showcases serigraphed stars and constellations on its inner surface, with a small Earth globe mounted inside. This fascinating object served as an instructional tool for pilots, sailors and astronauts to navigate using the stars. The globe's striking twentieth-century design also made it a prop for an iconic NASA photo shoot featuring astronaut John Glenn [see page 134].

CRAM

Veteran Map Maker

George Franklin Cram (1842-1928) served under General Grant during the American Civil War and participated in the renowned March to the Sea. Upon returning to Chicago, he assumed control of his uncle Rufus Blanchard's (1821-1904) map business, which had published the first *Chicago Street Guide*. He renamed it the George F. Cram Company, and eventually became the number one U.S. globe maker of the twentieth century. Cram sold his company in 1921 to Edward A. Peterson (1882-1966), the owner of the National Map Company. Peterson merged the two businesses under one brand, known as Cram. Peterson's commercial approach was to produce affordable globes for a broad public, making Cram one of the first globe makers to adopt such a strategy in the twentieth century. Cram himself never saw a globe bearing his name, as globe production only began four years after his death [no. 2].

Popular Pricing

Cram's moved to Indianapolis in the 1930s. They crafted remarkable models such as the Official R.E. Byrd Globe, in honor of R.E. Byrd (1888-1957), the aviator who made the first flight over the North Pole and later explored the Antarctic. The Sun Ray Globe was a demonstration instrument for the day and night cycle [no. 67]. Additionally, Cram also fulfilled custom orders, including globes adorned with slogans for various companies selling products as diverse as ties and heaters, and bank globes for financial institutes [no. 172]. The Colored Ocean Globe was a successful attempt to introduce decorative globes with a contemporary touch [no. 27]. The company also published illuminated glass globes mounted on bronze sculptures, such as the Dolphin Globe or the Atlas Globe [no. 28]. During the Second World War, Cram addressed the problem of rapidly changing geopolitical boundaries by offering Self-Revising Globes.

Expansionist Policies

Forty percent of Cram's revenue came from retail sales, with the remaining sixty percent from educational sales. There were School Globes for Beginners, Middle Grades, and Upper Grades, each increasing in complexity, adapted to the age group. Various stands were available, including the popular Cradle Globe Mounting. Additional products included Blank Slate Globes, Project Globes and celestial globes [nos. 30, 63, 179]. In 1966, Loren B. Douthit (1909-1996), a longtime employee at Cram, acquired the company. Together with his two sons, he pursued an expansionist policy, achieving an annual output of half a million globes and expanding their export to twenty-five countries. Cram acquired other companies to drive innovation. In the 1990s, it introduced the first Digitial-Vacuum-Formed-Illuminated Globe. In 2005, the family sold the company to publisher Herff Jones, which had previously acquired Nystrom, but subsequently ceased production of Cram's globes.

Cram's Outer Space and World Globe Handbook
Cram's /
Indianapolis, 1962

165

The George F. Cram Company from Indianapolis produced this school globe with a physical map. The model was available in both twelve-inch and sixteen-inch sizes. The metal Cradle Globe Mounting featuring a graded horizon ring could be used in two ways, either up or down. The sphere was called Tuffy Globe:
"Tuffy may be hammered, dropped and bounced without ill effect. Pittsburg Testing Laboratories have found this ball will hold up to a ton without breaking ... unbreakable when falling from the second floor ...."
Additionally, it had a Markable-Kleenable finish, allowing students to write on and erase its surface.

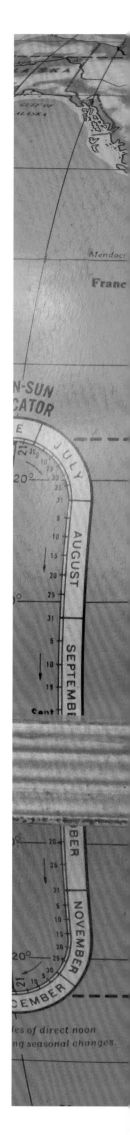

REPLOGLE

GLOBES

A Globe in Every House!

Luther I. Replogle (1902-1981) worked as a globe seller at Weber Costello in Chicago. When he lost his job during the Great Depression, his wife Elizabeth McIlvaine (1900-1937) began building globes in their basement while Luther sold them out of the trunk of his Model T Ford. Replogle's vision was to popularize globes, adopting the slogan *"A globe in every house!"*. His breakthrough came with a contract for one hundred thousand souvenir globes for the 1933 Chicago World Fair. Replogle invested early in technologies to produce globes more quickly and efficiently, such as steam-pressed cardboard spheres. He designed popular streamlined globes to fit the interiors of people's living rooms [nos. 26, 38]. Thanks to his innovative approach, the company became one of the largest U.S. globe producers of the twentieth century and remains in business today.

A Globe of Their Own!

In the postwar period, Replogle tapped into the potential of the children's market. The idea was to encourage parents to buy a globe for every child: *"A globe at their own ... makes homework easier ... helps them get better marks ... inspires them to become good citizens."* They introduced a line of colorful game globes that promoted the combination of education and fun [nos. 13, 115, 131-133]: *"To touch Rome or London, to put a finger on a mountain - The gift that inspires children to learn ... their own world to explore!"* To keep children engaged, Replogle published the Talking Globe, which came with a record narrated by world traveler Lowell Thomas Jr. (1923-2016). During the Space Race, Replogle created many globes aimed at children, such as the Satellite Globe, featuring a wooden ball on a metal ring to simulate satellite orbits, and various celestial, Moon, and Mars globes [nos. 43, 45, 103, 111, 166].

Mergers and Buyouts

Replogle published more than a hundred different models in many sizes and materials [nos. 68, 92, 170, 173, 174, 177]. The company was an early adopter of plastic. Initially, it glued paper map gores onto plastic spheres, but later transitioned to printing and pressing full plastic globes. In 1959, Luther I. Replogle sold his company to Meredith Publishing but remained as CEO. To expand into Europe, he founded the Danish joint venture Scan-Globe. After retiring, he was appointed ambassador to Iceland. In 1973, longtime employee William C. Nickels (1921-2002) bought the company from Meredith Publishing. The Broadview plant manufactured one million globes a year, accounting for two thirds of the world's production. In 2010, the Nickels family sold the company to Herff Jones. Four years later, another employee buyout ensured Replogle's position as one of the world's largest independent globe makers today.

The World's Only Talking Globe - The Story of Mr. World
Vinyl Record /
Replogle /
Chicago, 1950s

166
Replogle, in Chicago, produced this small celestial globe, called The Little Dipper in the 1950s. Its six-inch metal sphere features a drawing by Replogle's chief cartographer Gustav Karl Brueckmann (1886-1962). He depicted the constellations as classical figures in blue duotone, along with interconnected stars. The sphere swings within a full meridian on a steel wire stand. The base holds an explanatory booklet. A similar model, called Surprise Globe, was a terrestrial globe with hinged hemispheres. When opened, it revealed a hollow celestial map inside, accurately representing the concave universe [NO. 131].

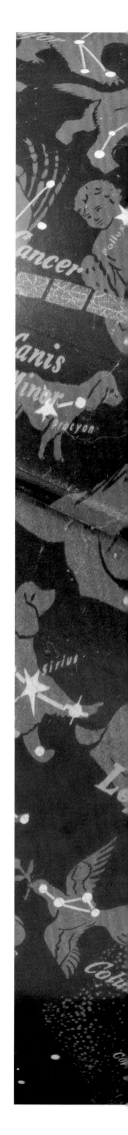

WEBER COSTELLO

Chalk and Wipers

C.F. Weber (+1913) took over the globe division of school supplier A.H. Andrews & Co. under its own name. The company used various map gores, some imported from British globe makers and some manufactured in-house. After a fire, the company built a new plant in Chicago Heights. Vice-president Thomas H. Costello (1854-1926) took over in 1909 and renamed the company Weber Costello Company. The firm, renowned for its blackboards, chalk, and wipers, became one of the largest U.S. school suppliers and a dominant globe maker [no. 171]. With much fanfare, Weber Costello introduced the Peerless Globe in the 1920s: *"The manufacture of globes requires skilled, experienced workmen, and the most accurate machinery. We have for this the largest and best-appointed establishment in the world. There are no globes made anywhere, by anyone, that is equal to our product."*

Churchill and Roosevelt

Weber Costello produced affordable school globes, featuring simple maps, mounted on steel wire bases [no. 46]. In the 1930s, the company introduced art deco-styled models, with streamlined metal stands or even airplane-shaped cast-metal bases [no. 39]. They also created custom-made globes for advertising purposes [no. 72]. During the heat of the war, Weber Costello built two fifty-inch globes, one for Winston Churchill (1874-1965) and one for Franklin D. Roosevelt (1882-1945), as a Christmas gift from General George C. Marshall (1880-1959). Marshall wrote: *"Today the enemy faces our powerful companionship in arms which dooms his hopes and guarantees our victory. In order that the great leaders of this crusade may better follow the road to victory, ... so that you may accurately chart the progress of the global struggle of 1943 to free the world of terror and bondage."*

Blackboards and Whiteboards

In the postwar period, Weber Costello introduced many new globe models with complex names, such as the Reality Political Physical Globe, the Tri-Graphic Educator Contour Relief Globe, the Semi-Contour Political-Physical Globe and the True Vue Globe [nos. 11, 175]. They ranged from inexpensive school globes to luxury models priced at thousands of dollars, like the Aristocrat Globe, designed *"... for the library, executive office, financial institution or similar location...."* Despite these innovations, Weber Costello's main activity remained blackboards, chalk, and erasers. In the 1960s, school supplier Beckley Cardy bought the company. Over time, traditional educational tools such as globes and chalkboards began to fade away, and production eventually ended. Reflecting the changing times, the company was taken over in the 1990s by Quartet Manufacturing, a specialist in whiteboards.

Around the World
Globe Manual /
Weber Costello
Company /
Chicago, 1958

167
The Weber Costello Company made this twelve-inch Moon globe during the Space Race. The globe's cartography consists of thousands of shaded moon craters, with a dense distribution of official names. Landing sites of the American Rangers and Surveyors are indicated, but not those of the Soviet Lunas. The Apollo landing sites are shown up to Apollo 14, dating this edition to 1971. The Apollo 11 landing site is hard to find, as it is located on the Moon's equator, obscured by the seam of the globe. The cradle base has a distinctive four-toed shape, made from a plastic wood imitation.

RAND MCNALLY

GLOBES

Train Tables and Road Numbers

William H. Rand (1828-1915) started a printshop in Chicago in the nineteenth century and partnered with the young Andrew McNally (1836-1904). Their breakthrough came with a contract to print *The Chicago Tribune* and the railroad's maps and timetables. To print colored maps, Rand McNally pioneered wax engraving. The company grew, and commissioned architect Daniel Burnham (1846-1912) to design a new building. It was the first high-rise to use a steel structure, igniting the skyscraper race. When automobiles conquered America in the 1910s, Rand McNally expanded into road maps, with its *U.S. Road Atlas* becoming a classic. In order to commercialize the maps, the company initiated the road number system by placing numbered poles along the roads. With a staff of nine hundred, it was the largest American map publisher of the early twentieth century.

Gypsy and Nomad

Rand McNally began producing globes at the end of the nineteenth century. In the 1920s and 1930s, it offered a broad range of models, including a sixteen-inch physical globe designed by Paul J. Goode (1862-1932), the cartographer of the standard *Goode's World Atlas*. In the 1940s, Rand McNally responded to the aviation era with the Air Age Globe [no. 71]. The 1950s saw many elegant models for the office or living room, with evocative names such as The Crusader, The Gypsy and The Nomad. Rand McNally also produced school globes. The Graded Maps and Globes Program offered five different-sized globes, each with a higher complexity of geographical information, to address various levels of the school curriculum. A cartographic innovation was the scribing process, where lines are scratched onto a translucent film to create a fine negative map as the basis for a sharp print.

Museum Spinners

Like most globe makers, Rand McNally participated in the Space Race with a Celestial Globe, a Satellite Globe and a Lunar Globe [nos. 94, 109, 168]. At NASA's request, the company constructed eight geophysical globes, built of seventy-five-inch diameter spheres made of epoxy reinforced fiberglass. The giant globes, featuring the Earth's natural colors and raised relief, were used to train astronauts to observe Earth as seen from their spacecraft. Rand McNally produced similar motorized monumental globes for museums and television studios, which could spin for demonstrations [see page 122]. The company was owned by the McNally family for five generations. Through many takeovers, the company grew to a staff of four thousand by the 1980s. At that time, globe making was abandoned, but Rand McNally remains an important American road map maker today.

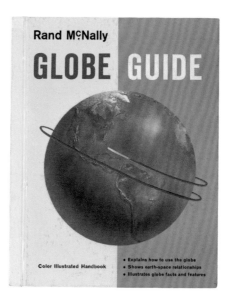

Globe Guide
Manual for a terrestrial globe /
Rand McNally /
Chicago, 1960

168
Rand McNally, the Chicago road map publisher and globe maker, was an early adaptor of Bakelite globe stands. It created various models in the 1940s, such as this ten-inch celestial globe, called The Galileo. It has a star map by astronomer Oliver J. Lee (1881-1964). The stand and horizon ring are made of Bakelite, cast as one single volume. To make the brown material more appealing to the public, the company marketed it as an expensive wood: *"The warm tones of the globe bleed in with the rich walnut-finished Bakelite stand and harmonize with any surroundings in which the globe is placed."*

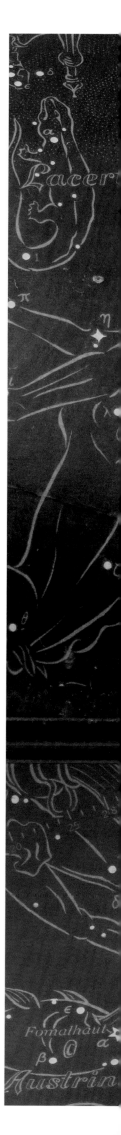

More American Globes

169

The Rand McNally company created this twelve-inch globe in the 1950s. It has detailed cartography, with sea and air routes. The sphere is hinged to a gimbal meridian. It sits on a steel wire stand, which is bent in a curve, giving the globe a lightweight, cheerful character, radiating the optimism of the postwar period.

170

This twelve-inch globe, created by Replogle in Chicago in the 1940s, features a black ocean map. The origin of the remarkable stand, a triangle of bent copper wire, is less certain. As no similar globes have surfaced, it seems to be a one of a kind, possibly a prototype or a replacement of the original stand by an amateur.

171

Weber Costello Company in Chicago patented this globe in the 1910s. The twelve-inch plaster sphere, covered with paper map gores, balances on the tip of an inclined wooden stand. Its asymmetrical composition of oak wood is crafted in the mission style, a style used in the early 1900s by the American arts & crafts movement.

172

Globe maker Cram of Indianapolis customized a standard seven-inch model into an advertisement globe in the 1930s for The Buffalo Evening News. It was achieved by simply glueing a round piece of printed paper with a publicity text onto the metal base. The newspaper offered the globe for 39 cents as a bonus for new subscribers.

173

Replogle in Chicago offered this ten-inch Air-Ways-Globe as an aid for air travelers during the aviation boom of the 1940s. The globe depicts the network of air routes with red lines. A scale is provided to measure circle distances and flying times. To save on materials during war time, the cradle base is made of pressed cardboard.

174

In the 1960s, the National Geographic Society offered globes to its magazine readers, made by Replogle. The twelve-inch globe rests on a transparent Lucite cradle base, equipped with a horizon ring for calculating time and distance, as well as a hemispherical transparent dome, a so-called geometer, to measure Earth's surfaces.

More American Globes

175

Weber Costello, a school supplier from Chicago, made traditional and modernist-styled globes in the 1950s. This model has map gores with brightly colored cartography. The twelve-inch sphere is supported by a remarkable base, shaped in the form of two wooden wedges, with the sphere fixed by means of a cantilevered metal pin.

176

This globe was made by Rand McNally in Chicago in the 1960s. The paper map gores show a black ocean map on a twelve-inch sphere. The stand is made of steel wire, bent in the shape of a booklet holder, to accommodate the accompanying manual. It is a lightweight version of the 1930s atlas globes, which had a wooden bookcase with an atlas for a base.

177

Replogle of Chicago manufactured this interesting globe during the Second World War in the 1940s. The cartography shows the British Asian colonies during wartime. A full meridian encircles the twelve-inch sphere, which is fixed to a glass base. The choice of a glass base was due to the scarcity of metal, which was reserved for the war industry.

178

Denoyer-Geppert, a Chicago school supplier, launched a series of cradle-based globes in the 1940s, aimed at schools. This model, The Liberty, features a twelve-inch sphere on a wooden cradle. The robust orb, called the Cartocraft Beginners Globe, displays simplified cartography and large letters. It was advertised as unbreakable.

179

Cram from Indianapolis marketed Project Globes in the 1950s. This model is a sixteen-inch globe resting on a metal cradle base called Clear View Mounting, because one can observe the sphere without obstruction. The cartography indicates no details or names, only the landmasses and the lines of the equator and the Tropic of Cancer and the Tropic of Capricorn.

180

Philip & Son, a British map and globe maker, introduced this contemporary designed globe in the 1950s. It has a forty-eight-centimeter plastic sphere with glued-on paper map gores. The heavy cast-metal base has four feet that form a cradle. However, since the sphere is both illuminated and fixed to a transparent full meridian, it cannot be picked up.

FU-
TURE

A World in Your Pocket

At the close of the twentieth century, the world began to shrink. Distant places became intertwined by the intense exchange of trade, culture and knowledge. The development of the internet in the 1990s accelerated the globalization process, connecting people across the globe.

In 1976, a young Steve Jobs (1955-2011) built the first successful personal computer in his Californian garage and co-founded Apple Computer Inc. It would later become the world's largest company of the twenty-first century, renowned for its many disruptive innovations. Thirty years later, on January 9, 2007, a now grey-haired Steve Jobs, clad in his iconic black turtleneck, took the stage in San Francisco. With a frail voice he uttered the historic words: *"… This is the day that I have been looking forward to …."* That day he unveiled the iPhone, a revolutionary concept that combined a computer, telephone, and a multimedia platform into a compact, hand-held device with a smart design and a user-friendly interface. The smartphone drastically changed the way we communicate and access information.

Not long after, everybody was walking around with an accurate digital globe of the Earth in their pocket, containing infinitely more geographical information than any globe before. The good old analog globe was demoted to an object of decoration, ushering in the new era of digital globes.

During many centuries, globes have served as spherical scale models to represent the Earth. In the twentieth century, the classical handcrafted globe was replaced by mass-produced generic plastic globes. Simultaneously, efforts were made to create innovative representations of the Earth and the planets, such as folded globes, photographic globes, and 3D printed globes. At the dawn of the twenty-first century, a digital globe has found its way into everyone's pocket.

FOLDABLE

Cardboard and Supermarkets

Robert Gair (1839-1927), was a Scottish paper maker in New York. One day, he observed a press accidentally folding cardboard instead of cutting it. Seizing the opportunity, Gair decided to develop machine-folded cardboard boxes. It became so successful, that he built a new factory in Brooklyn in 1904. The iconic Gair Building, made of reinforced concrete, revolutionized architecture in the same way cardboard boxes revolutionized food distribution. Small portions of food were prepackaged in colorful cardboard boxes, which customers could grab from open shelves and pay for at checkout in the first supermarkets, which opened in the 1910s. Similarly, a folded cardboard globe presented a viable alternative for a heavy traditional globe. Twentieth-century architects took up the challenge to fold a single piece of flat printed cardboard into a globe.

Butterfly Map

Scottish architect Bernard J.S. Cahill (1866-1944), who immigrated to San Francisco, made a radical attempt to create a globe from a single piece of cardboard. Cahill's so-called Butterfly Map consists of eight rounded-off triangles that together form a butterfly shape. When folded and glued, the octahedron approximates a sphere. The advantages include efficiency of production, and the fact that the unfolded map is less distorted than a traditional Mercator projection. Cahill developed a number of variations of the design, before patenting his folded cardboard globe in 1913. He also patented another model, which was printed on a piece of rubber with incisions, allowing it to be turned into a rubber globe. Cahill founded the Cahill World Map Company, but his folded globe was never taken into production. He custom-made Butterfly Maps with a thematic cartography, such as a special edition commissioned by the Port of San Francisco [no. 194].

Dymaxion World

Richard Buckminster Fuller (1895-1983), an enterprising American engineer, patented a folded cardboard globe in 1946, known as the Dymaxion World. His mathematical design consisted of a flat map featuring eight triangles and six squares, forming what is known as a cubo-octahedron. When assembled, this structure approximates a sphere [nrs 181, 192]. Similar to Cahill's map, Buckminster Fuller's unfolded map is less distorted than a traditional Mercator map. A remarkable feature of the Dymaxion World Globe is its ability to be unfolded into many different configurations, as a sort of geographic origami. Each configuration arranges the continents in different spatial relations, creating different thematic maps that alter our perception of the world. Buckminster Fuller used similar polyhedrons to create spherical buildings, made of lightweight geodesic domes.

Dymaxion World
Richard Buckminster Fuller holding his cubo-octahedron foldable globe / *Philadelphia, 1950s*

181

The Dymaxion World, designed by Richard Buckminster Fuller and patented in 1946, is a cardboard globe. The flat printed map can be folded into a so-called cubo-octahedron, a three-dimensional form consisting of eight triangles and six squares, which approximate a sphere. The longitudes and latitudes on each of the globe's fourteen sides are almost undistorted, offering the advantage that the unfolded flat map is more accurate than a Mercator projection. The flat map can be configured in various ways to create thematic maps of the world, each presenting different geographical relationships.

CUT-OUT
GLOBES

Efficiency Movement

Frederick W. Taylor (1856-1915) was a foundational figure in the Efficiency Movement, also known as Taylorism, during the early twentieth century in America. His goal as an engineer was to achieve a more prosperous society, by eliminating waste across industrial, economic and social realms through the 'scientific management' of engineering and organizational processes. In the globe business, similar efforts towards efficiency were also pursued. Many skills and materials are necessary to produce the heavy, bulky and hard-to-transport objects. As a result, several cost-effective and convenient alternatives were developed, such as globes printed on a piece of cardboard that could be mailed, cut out, and assembled by the customer. As early as 1913, Alois Höfler (1853-1922), a Viennese professor of education, designed a convincing cardboard cut-out celestial globe [no. 191].

Assembled in Five Minutes!

Ten years later, Alfred J. Nystrom patented the Form-a-Globe. For twenty-five cents, this Chicago school supplier offered students a package of flat cardboard map gores that were attached at the equator but loose at the poles. To assemble it, one had to connect the gores at the tips with a steel wire, creating a sort of paper lantern. After the geography lesson, the globe could be flattened again and carried in a school bag. Nystrom marketed it with the slogan: *"Assembled in Five Minutes!"* [no. 193]. A similar globe was published by Georg Westermann in Germany, the Schüler-Globus zum Selbstformen. In Madrid, publisher Saturnino Calleja (1853-1915), renowned for his illustrated children's books and cardboard construction plates, released a flat cardboard cut-and-paste Armillary Sphere. Its design resembles a composition by the Russian constructivists [see image above].

Scarce Materials

To circumvent material shortages during and after the war, alternative materials were utilized in globe production, such as cut-out cardboard globes. Harry M. Gousha (1892-1970), who had worked for Rand McNally before starting his own business in Chicago, was one of the major U.S. road map makers, a product that involved complex folding technology. In the early 1940s, Gousha released the Plano-Sphere, a flat cardboard globe. Two cut-out color-printed cardboard hemispheres slide together to form a cross that stands upright on a table. The maps on opposite quadrants create an optical illusion of a globe. This also served as an attractive display for advertisements. C.R. Anthony Company, a chain of family-owned department stores in the U.S., utilized a similar cut-out cardboard globe in the 1950s, to promote its *"68 department stores serving the Southwest"*.

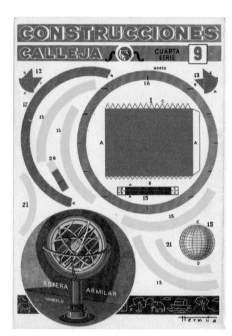

Esfera Armilar
Cut-out cardboard
Armillary Sphere
by Saturnino
Calleja / *Madrid,
1940s*

182
Gerald A. Eddy (1889-1967), an American cartographer of tourist maps, patented this Fold-O-Globe in 1942. It consists of two color-printed cardboards, stapled together. When unfolded, it creates the illusion of a globe. Eddy promoted it with aplomb: *"... Round like the Earth, this Fold-O-Globe drawn in perspective presents the first important innovation in map making in 400 years ... a continuous map projection showing at a glance the 'Countries of the World and their Flags'. Note the adjusted equator, which gives the illusion of Earth's curvature and the true relation of one continent to another...."*

FICTIONAL MARS

GLOBES

Martians in New Jersey

American actor and movie director Orson Welles (1915-1985) broadcast a radio play in 1938, a common format in a period when television did not yet exist. The *War of Worlds* as it was called, was based on the novel by British science fiction author H.G. Wells (1866-1946), one of the earliest stories about an encounter between humans and extraterrestrials. Welles staged his report about an invasion of Martians in New Jersey. The radio broadcast sounded so realistic that it caused mass hysteria: listeners believed that a real alien attack was unfolding, and public panic ensued. The fascination with Mars and the fear that Martians were about to invade the Earth with their flying saucers never faded away. Its origins can be traced back to false theories of an inhabited red planet, which spread at the beginning of the twentieth century.

Mars Men
Orson Welles and
H.G. Wells studying
a black ocean
terrestrial globe /
1940

Life on Mars

Milanese astronomer Giovanni Schiaparelli (1835-1910) sparked Mars mania. In 1877, he observed straight lines on the planet and created a new Mars map. French astronomer Camille Flammarion (1842-1925) in his book *La Planète Mars et ses conditions d'habitabilité (The Planet Mars and Its Habitability Conditions)* further popularized the myth that these lines were canals made by Martians. The idea was hugely successful and American amateur astronomer Percival Lowell (1855-1916) took it even further. He claimed the canals were dug by a highly developed civilization to direct water from the poles to irrigate the equator. This assertion triggered a feverish Mars mania among the public and inspired early twentieth-century globe makers to create canal covered Mars globes. The fear of Martians never went away. It took decades to prove that the Martian canals were just an optical illusion.

Flying Saucers

In 1947, an Unidentified Flying Object (UFO) crashed in the New Mexico desert near Roswell. According to conspiracy theorists, authorities salvaged the corpses of extraterrestrials and kept it a secret. Since then, there have been many public sightings of flying saucers, the supposed spacecraft of aliens. Speculations about Martians and their canals ended in the mid-1960s when the Soviet spacecraft Mars 1 and the American Mariner flew close to the red planet but found no evidence. Globe makers had to completely redesign their Mars maps. Based on blurry photographs sent back to Earth by the probes, LeRoy M. Tolman (1931-2015), chief cartographer of Replogle in Chicago, airbrushed a speculative map with a lot of artistic freedom in the late 1960s. This map was used for a fifteen-centimeter lithographed metal Mars globe [no. 45].

183
In 1957, Walt Disney (1901-1966) created an animated documentary called *Mars and Beyond*, at a time when Mars was still a mysterious planet surrounded by speculation. The movie includes psychedelic animations of bizarre Martian life forms. Disney may have been inspired by the surrealist paintings of Salvador Dali (1904-1989), with whom he collaborated in the 1940s. The unique ten-inch Mars globe above was used by Disney to create animations for the movie. It is a standard Replogle globe, hand painted with a map of Mars that still includes the false Martian canals.

SCIENTIFIC MARS

GLOBES

Mars Mosaic

Planetary cartographers Patricia M. Bridges (1934-) and Jay L. Inge (1943-2014) worked at the U.S. Geological Survey, in the Lowell Observatory of Flagstaff, Arizona. They drew the first U.S. Moon map in the 1960s, and then took on the challenge of compiling the first NASA Mars atlas. In the early 1970s, Mariner 9 orbited the red planet and transmitted detailed photo images of the entire surface back to Earth. After a hundred years of speculation, it was finally possible to draw a scientifically accurate map of Mars. The team reversed the traditional cartographic process: their first step was to create a forty-eight-inch mosaic globe by glueing fifteen hundred photo fragments together. This globe formed the basis for the drawing of a detailed map of Mars. In 1980, Replogle used a revised version, updated with data from the Viking missions, to create a twelve-inch Mars globe for NASA.

Airbrush and Adobe

Creating a Mars map involves knowledge of astronomy and geology, as well as artistry. Planetary map makers used an airbrush to spray-paint their maps, allowing the cartographer to create a suggestive image. As explained by Patricia M. Bridges: "... *the shaded relief of Mars should not be considered a copy of a photograph. The final map synthesizes several hundred photographs and clarifies dramatically the alien topographic structure of a planet....*" Seasonal elements, such as sandstorms or polar ice caps complicate matters, but the main challenge is a lack of contrast. Terrestrial maps have countries, landmasses, and oceans. Mars has neither, so other features are used, such as albedo, the light and dark patterns generated by reflections of different soil types or shadows of craters, to create a 3D effect. The color is usually red ocher, referring to the planet's ferrous soil.

Venus and Mars

The Soviet Union had an early success in 1963 with the first flyby of the uncrewed spacecraft Mars 1 Eight years later, Mars 2 went into orbit around the planet, two weeks after the American Mariner 9. In the same year, Mars 3 made a first soft landing on the planet's surface. The Soviets compiled various Mars maps, but it wasn't until 1989 that the Sternberg Astronomical Institute in Moscow published the first Soviet Mars globe, an edition of ten thousand copies. This particular Mars globe features a relief drawing by V.I. Stutshnova, based on the Mars map of the U.S. Geological Survey. The paper map gores, printed in an overall reddish-yellowish color, are affixed to a plastic sphere mounted on a white plastic tripod stand. Several Soviet planetary globes followed in the years after, including a Venus globe and a Phoebus globe, one of Mars' moons.

Photo Mosaic
Mars Globe
Scientist compiling
48-inch Mars
globes /
NASA Jet
Propulsion
Laboratory /
U.S., 1971

184
Denoyer-Geppert produced this Visual
Relief Mariner 9 Mars Globe in 1973. It is
based on the photographic data sent to
Earth a year earlier by the American Mariner
9, the first spacecraft to orbit another planet.
The artistic interpretation of the paper map
gores was created by planetary cartographer
Jay L. Inge, enhanced with a more intense
red-brown color to represent Mars.
The sixteen-inch globe suggests that
the planet is very hot, although in reality,
it is quite the opposite. The globe rests
on a sober plywood cradle base, varnished
in the same red color.

ART

International Klein Blue

French artist Yves Klein (1928-1962), born in Nice, began his career as a black belt judoka. His next step was to move to Paris and dedicate his life to art. Klein created monochromes in an effort to *"liberate color from the prison that is the line"* and *"make visible the absolute"*. In his quest to capture immateriality and the infinite, he developed a special blue color, a new formula branded International Klein Blue (IKB). It is a bluer-than-blue color that radiates vibrant waves, engaging the eyes of the viewer, *"... allowing us to see with our souls, to read with our imaginations"*. Klein applied it to many works, including on an IKB-globe. He wrote the following: *"In 1957 Yves Klein stated that the Earth was entirely blue. It was on that occasion that he created a Blue Relief Globe ... Four years later cosmonaut Gagarin stated in April 1961 that the Earth is of a deep intense blue!!!"* [see page 246].

Calfskin

Italian artist Claudio Parmiggiani (1943-) worked in the spirit of the ZERO movement. Several of his works include maps and globes, depicting Earth as a vulnerable organism. One work features an inflatable globe, folded in a closed glass jar, marked with a label 'Globo', suggesting that Earth is fragile and needs protection. Another work called *Terra* is a large, seventy-centimeter terracotta globe. During a performance in 1989, Parmiggiani buried the globe in the garden of the Museum of Contemporary Art in Lyon. The idea is that the Earth is invisible as a whole, a precious object buried like a treasure: *"... a sculpture that wants to be secret, invisible, born for no exposition and for no public ..."* [see picture above]. In his work *Tavole Zoogeografiche* he manipulates pictures of five cows, giving each one a skin pattern in the form of a map of one of the continents.

World Processor

German artist Ingo Günther (1957-) has been working on his *World Processor* for forty years already. The aim of this project is to map the global condition by using illuminated globes. Günther has created more than a thousand thematic globes based on objective data. These topics range from political conflicts and social issues to the natural world and other concerns. The data are immediately comprehensive and intuitively accessible, thanks to their strong visual presentation. The colorful globes are exhibited in monumental installations, creating a memorable experience that encourages visitors to engage with the themes and with each other. In Günther's own words: *"... I am challenged by the incomprehensibility of the world's totality ... every globe requires the invention of a new code in order to represent the data appropriately and effectively ..."* [no. 198].

Terra
Performance by artist Claudio Parmiggiani / Musée des Beaux-Arts / *Lyon, 1989*

185
Italian artist Claudio Parmiggiani worked in the spirit of the ZERO movement. Several of his artworks incorporate a globe, such as this one from 1968, called *Pellemondo* (Skinworld). The sphere is enveloped in calfskin and mounted on an aluminum stand. The distinctive black-and-white pattern of the calfskin resembles a world map, with black landmasses contrasting against a white ocean. However, the tactile surface of the hairy animal skin suggests that the world is a precious living organism, one that must be treated with respect, nurtured, and protected by humans.

"In 1957 Yves Klein stated that the Earth was entirely blue. It was on that occasion that he created a Blue Relief Globe ... Four years later cosmonaut Gagarin stated in April 1961 that the Earth is of a deep intense blue!!!"

Yves Klein, 1961
French Artist

Yves Klein
and his Blue
Relief Globe (RP7) /
1961

PHOTOGRAPHIC
GLOBES

Aerocartography

Gaspard-Félix Tournachon (1820-1910), a pioneer in photography and ballooning in France, is better known as Nadar. He took his camera on board and started aerophotography in the 1850s. During the First World War, the British Army mapped the network of German trenches by capturing images from small airplanes, establishing a new method of map making. Forty years later, the Americans developed the U-2 spy plane, capable of flying unnoticed at a height of twenty-one kilometers, to capture detailed reconnaissance photographs. The first picture of Earth from outer space was captured at an altitude of one hundred and five kilometers. It was taken by a camera inside a V-2 rocket launched from New Mexico in 1946, which had been seized by the Americans from the Germans. By the end of the century, hundreds of satellites orbited the Earth, capturing images of its surface from space.

Blue Marble in Black Universe

On Christmas Eve 1968, millions of people on Earth were glued to their TV sets, eagerly following the first men to fly around the Moon. To mark the drama of the moment, the astronauts aboard Apollo 8, Borman, Lovell and Anders, recited aloud the opening verses of Genesis: the Creation of the Earth. As their spacecraft turned around the Moon and redirected itself towards Earth, the astronauts were stunned. They became the first people in history to witness Earth floating like a splendid blue marble in the vast expanse of a black universe [no. 1]. In that very moment, the image of the Earth as a whole was finally captured, the very image that globe makers had sought to depict for more than five centuries. The ultimate result would be a photographic globe. Replogle embraced the challenge and produced Planet Earth, a twelve-inch globe crafted from photographs taken from space [no. 186].

Overview Effect

A photographic globe of the Earth seems intriguing but upon closer inspection, it proves to be somewhat disappointing. The man-made world of borders, cities, railroads, shipping lines, time zones, and so on is not visible from space. It is precisely these elements that people are eagerly seeking on a globe. Consequently, few photographic globes have been produced. One notable exception is a seven-meter-diameter inflatable globe created by British artist Luke Jerram (1974-), a touring artwork called *Gaia*. Based on a photograph taken by Apollo 17, it rotates slowly in the air. The public experiences Earth as an astronaut would: "...*Gaia creates a sense of the overview effect, the experience of awe for the planet, a profound understanding of the interconnection of all life, and a renewed sense of responsibility for taking care of the environment, as felt by astronauts in outer space...*" [no. 195].

Mapping enemy positions
A military aerial photographer at work during the First World War / 1914-1918

186
Scan-Globe, the Danish subsidiary of Replogle, released the Planet Earth Globe in the late 1980s, a twelve-inch illuminated plastic sphere. It presents what would appear to be a photographic image of Earth, but it is actually a rendering by Boston design office Krent/Paffett/Teifert, based on a satellite image. Marketed with the slogan *"A View Few Witness"*, the globe portrays Earth as seen from space, with blue oceans, green forests, and yellow deserts. Clouds are depicted only above the oceans, while rivers are widened to enhance visibility. It underscores the notion that geography is an abstraction conveyed through simplification and exaggeration.

MAGNETIC GLOBES

World Wide Web

The foundation of the internet was established in CERN in Geneva around 1990. Belgian Robert Cailliau (1947-) and British Tim Berners-Lee (1955-) proposed 'hypertext' to connect researchers at the European Laboratory for Particle Physics. CERN is home to the largest electromagnet on Earth, a twenty-seven-kilometer-long underground ring designed to accelerate elementary particles to the speed of light for the observation of physical phenomena. Earth itself possesses a natural magnetic field, which is used by ships as a navigation aid. This magnetism is produced by the continuous movement of hot liquid iron within the Earth's core. As a result, the magnetic poles shift their position by approximately fifty kilometers each year and do not align precisely with the geographical poles. Because of its dynamic character, the Earth's magnetic field is rarely depicted on globes.

Magnetic Levitation

In 1920, Danish educator Einar Georg Schlätzer patented a magnetic globe with an electrical coil inside to generate a magnetic field. When a compass needle was held close to it, it would align itself with the globe's magnetic pole. Unfortunately, this concept was never brought to production. In the 1950s, Replogle introduced the Magnetic Air Race Globe. Its name alludes to the magnetic pawns that are attracted to the metal game globe [no. 133]. To celebrate the turn of the millennium, Columbus Verlag created a magnetic levitation globe. It hovers in mid-air, unsupported, held aloft solely by a magnetic field, called Schwebeglobus [no. 187]. Over the years, the technology of the electromagnet was improved by other globe makers. Several levitation globes are on the market today, which only consist of a sphere floating in mid-air above a simple base, without any additional elements [nos. 199, 200].

Autorotation

Bill French (1934-), an American physicist, invented a novel type of globe in the 1990s. This small sphere rotates continuously around its axis without the need for electric wires or batteries. It is propelled by both the Sun and the Earth: integrated solar cells provide power and enable the globe to spin in response to the Earth's magnetic field. The outer layer of the globe is crafted from transparent acrylic, while inside, a secondary plastic sphere rotates as long as the globe is exposed to light. The map image is a satellite photo, depicting natural colors and white clouds, devoid of graphics or text. It is a very convincing scale model of the Earth, representing the ultimate physical globe. MOVA, the Californian company founded by French, promotes the globe as *"a home decor product"*. It is offered with various cartographic images, as well as depictions of the planets of the solar system [no. 196].

Magnetic acceleration
A 7.2-meter-diameter magnet arrives at the CERN building site / *Geneva, 1955*

187
Columbus Verlag, the innovative German globe maker, released this magnetic levitation globe called Schwebeglobus, to commemorate the year 2000. The large bow-shaped silver-metallic stand generates a magnetic field when electricity is applied. The field suspends the thirty-centimeter sphere, which hovers in mid-air and slowly rotates. The glossy plastic sphere features basic cartography of silver landmasses and black oceans, a stark contrast to the detailed globes typically produced by Columbus Verlag. It is evident that the globe is intended solely as a decorative conversation piece in an office or a living room, designed to impress guests.

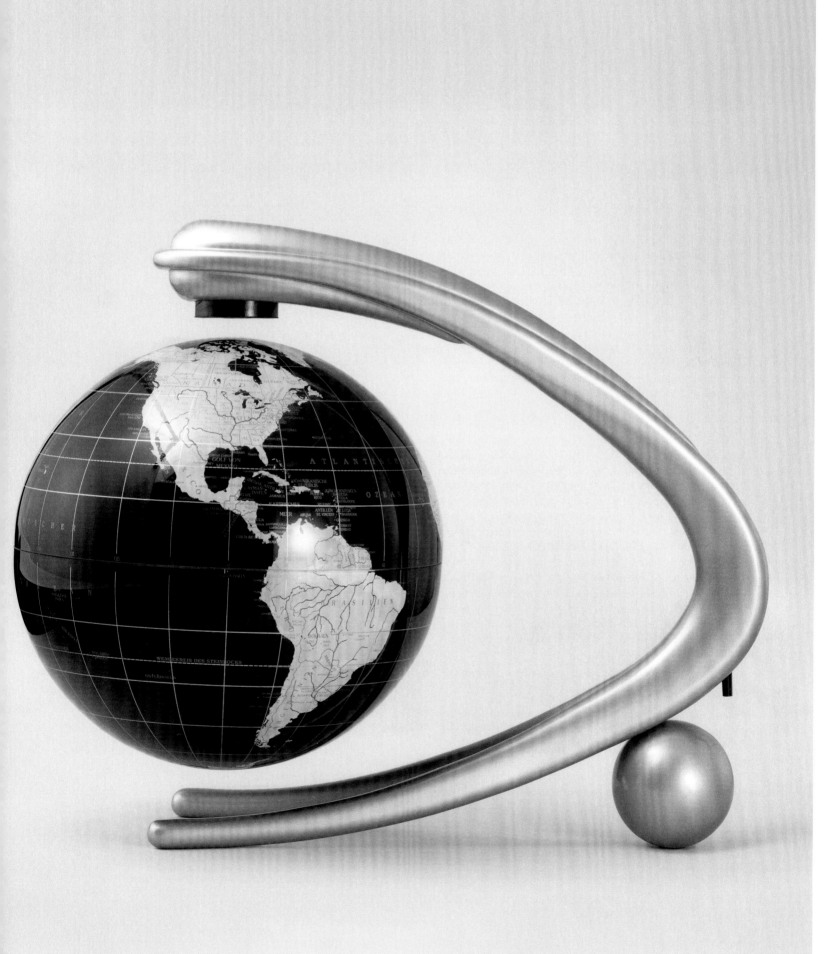

BIG DATA

GLOBES

Computer Database

English mathematician Alan Turing (1912-1954) laid the groundwork for computing in the 1930s. He conceptualized what would later be called a Turing machine, an imaginary device serving as a computational model. During the war, Turing was part of the British team that successfully deciphered the code of Enigma, Nazi Germany's secret messaging machine. After the war, Turing built the first experimental computers. He lived a short and tragic life and was prosecuted for his homosexuality, considered a crime in 1950s Britain. By the end of the century, computers were integrated in all aspects of live. Enormous amounts of digital information can be stored today. Cartographic institutes and commercial map makers set up geographical databases, which can be filtered to compile various maps for globes, or even to produce specific globes on demand.

Globes on Demand

The Commission for the Geological Map of the World has existed since the late nineteenth century, with the aim to develop Earth sciences maps. In 2018, it created the Earth Geological Globe, showcasing the latest advancements in our understanding of the planet's geology. The globe, produced by Replogle in Chicago, is compiled by digitalizing and filtering the most recent geological data provided by numerous international institutes. Contemporary technology enables the direct printing of the globe's map gores from a set of computer data. In this instance, it is a twelve-inch plastic globe with various colored surfaces representing different geological time periods. With this technology, any customized globe can be produced on demand, affordably and efficiently, as a unique piece or small series for a limited number of customers [no. 197].

3D Printers

The next evolution of globe manufacturing arrived in the twenty-first century. Today, a globe can be generated directly from a computer file by means of a 3D printer. New globe makers have emerged, such as Little Planet Factory in London, specializing in small 3D printed globes, using a computer-controlled process. The machine applies layer upon layer of synthetic material in various colors, until the desired globe is complete. Geometry, color, and relief are accurately printed onto a monolithic, seamless globe. The digital model can be updated any time with new geographic data. Instead of mass-producing millions of identical models, a single globe can be made on demand, at a low cost. After its long transformative journey through the twentieth century, globe manufacturing has returned to its origins: a unique object made by a skilled craftsman, which today we call a robot [no. 188, 200].

The Bombe
An early computer to decipher coded messages during World War II /
Bletchley Park U.K., 1940s

188
Little Planet Factory from London aspires to make all celestial bodies of the solar system available using 3D printing technology. This small twelve-centimeter globe depicts the birth of the universe. The map is created by the European Space Agency based on data gathered by the European Planck surveyor satellite in 2013. The colors represent the temperature differences in the cosmic background radiation of the universe, dating back 13,82 billion years, shortly after the Big Bang. 3D printing technology enables the direct production of a globe from a set of computer data.

GENERIC

Monobloc Chair

In 1942, a journalist wrote: "*Plastic furniture has appeared from time to time, but has been either too expensive or too fantastic in design to be practical. If these faults are overcome and a well-designed type of furniture is put on the market, the possibilities are enormous*". A few years later, Canadian designer Douglas C. Simpson (1916-1967) paved the way by developing a prototype called Monobloc chair, created through plastic injection into a single mold. French household companies, Allibert and Grosfillex, successfully brought a monobloc chair in production in the 1970s. Its nondescript design blends elements of a traditional wooden chair and a modernist plastic seat. Thanks to its lightweight construction, its stackability, and, above all, its low production costs of only $3.50, it became the ubiquitous chair of the century. As harbingers of globalization, billions of copies have proliferated across the globe.

Cheap and Unpretentious

The end of the century witnessed the global distribution of inexpensive, identical mass-produced goods. They lack a sense of time and place, and their design is an unpretentious combination of proven technology, low manufacturing costs, and fast production methods. Likewise, globes followed this trend, thanks to the advent of pressed plastic spheres technology. The 'generic globe' embodies an undefinable style, is lightweight and sturdy, affordable and accessible to all. It has evolved into a generic, mass-produced, globalized object: two plastic hemispheres, imprinted with basic cartography, glued together along the equator, fixed to a plastic half meridian and a plastic circular base. Each year, millions of these generic globes are manufactured and distributed worldwide. Among the largest producers are Tecnodidattica in Italy, Replogle in the U.S., and Fucashun in Taiwan [no. 189].

End of the Globe

Generic globes feature a reduced amount of geographical information. Compared to the detailed globes of the first half of the twentieth century, their role as a primary source of geographical information has diminished in the era of digital maps. The contemporary globe has become a metaphorical object. Just as every house may have a small garden as a comforting symbol of humanity's control over nature, so too does every home often contain a miniature model of the Earth, as a symbolic representation of humanity's mastery over the world. Man has conquered the world and diminished it. At the close of the twentieth century, after five centuries of transformation, the journey of the globe has reached its conclusion. Its final form is a generic plastic object, a decoration on a bookshelf, the perfect Christmas present, emitting its comforting glow in the darkness of children's bedrooms.

World Wide Chair
Man sitting on a generic white plastic Monobloc chair / *Nairobi, Kenya, 2011*

189
Tecnodidattica from Italy is one of the leading producers of the contemporary generic globe. This particular globe dates back to the 1980s. It consists of two printed plastic hemispheres joined at the equator. The globe rotates in a Plexiglas half meridian, set at the correct axial inclination of the Earth. The graded meridian is fixed to a circular plastic base. Illumination is a standard option. The cartography is adequate and straightforward. Generic models are available with maps of varying levels of detail and in different color schemes, catering to the various tastes of the customers.

DIGITAL
GLOBES

Smartphones

Steve Jobs (1955-2011) co-founded Apple Computer Inc. in his Californian garage. From the 1970s onward, Apple made many disruptive innovations and would become the world's largest company. On January 9th, 2007, clad in his iconic black turtleneck, the grey-haired Jobs took to the stage in San Francisco and uttered the now historic words, with a frail voice: *"...This is the day that I have been looking forward to ...."* That day he unveiled the iPhone, a revolutionary concept that combined a computer, a telephone, and a multimedia platform, housed in a small handheld device with a smart design and a user-friendly interface. The smartphone changed the way we communicate and access information. Not long after, everybody would be walking around with a digital globe in their pocket, containing infinitely more geographical information than any globe before.

Radar and GPS

Wireless intercontinental communication commenced at the beginning of the century. Radio stations were constructed, and radio beacons were installed on lighthouses to assist ships in locating their positions. In the 1930s, Radio Detection and Ranging (RADAR) technology was developed to detect ships and planes by emitting radio signals that reflect off metal objects and return to the sender. It played a decisive role in the outcome of the Second World War, with the British radar system outperforming the German counterpart. In the 1970s, the U.S. constructed the Global Positioning System (GPS), a navigation and surveillance system comprising satellites orbiting the Earth and emitting radio waves. The system continues to provide all data for maps and route planners. Today, cartographers create their maps based on data collected from space.

A World in Everyone's Pocket

Keyhole Earth Viewer, the first available digital globe, was launched in the 1990s. The entire map of the Earth was compressed onto a Compact Disc (CD), the data carrier of the late twentieth century. When inserted into a personal computer, users could view the image of the Earth, spin the virtual globe and zoom in to any desired level of detail. A budding internet browser company called Google discovered that many of its clients were searching for geographical information, such as maps or directions. In 2004, Google acquired Keyhole and transformed it into an internet-based platform called Google Earth [no. 190]. Shortly after Steve Jobs presented the iPhone, Google launched the application Google Maps. The twentieth century began with the ambition to have a physical globe in every house, but it ended with a digital globe in everyone's pocket. Mercator would have loved it!

iPhone displaying
Google Earth
application
2020s

190
Google, the IT company in Silicon Valley, California, developed Google Earth, a virtual globe accessible through an internet application. This virtual globe can be accessed and viewed on the screens of computers or smartphones. In contrast to the limited information provided by a physical globe, the geographical data available on a virtual globe, continuously updated by satellites orbiting the Earth, is vast. As a result, the traditional globe has lost its role as a source of geographical information and has been reduced to a decorative object. Today, a digital globe resides in everyone's pocket, serving as an endless source of detailed geographical knowledge.

More Folded Globes

191

Viennese professor of education Alois Höfler designed this cut-out celestial globe, published in 1913 by B.G. Teubner Verlag in Leipzig. One cardboard sheet has a print of the star map, to be cut by the customer into map gores, while the second sheet has a print of the stand and the horizon ring, also to be cut out and assembled.

192

Richard Buckminster Fuller, an enterprising American engineer, patented a folded cardboard globe in 1946, which he called the Dymaxion World. It consists of eight triangles and six squares, forming a so-called cubo-octahedron, approximating a sphere when folded. The globe can be unfolded into various configurations, akin to geographic origami. Each configuration alters the spatial relations of the continents and creates thematic maps. It is a powerful instrument, capable of transforming the observer's perception of the world. This image depicts a 1954 version of the flat unfolded Dymaxion Airocean World Map.

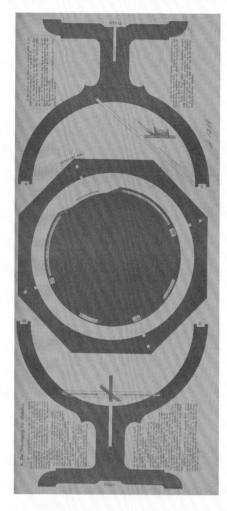
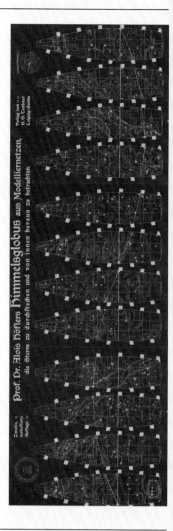

193

School supplier Nystrom of Chicago patented the Form-a-Globe in 1923, a cost-effective cutout cardboard globe designed for students. The map gores are affixed at the equator but left loose at the poles. They can be connected using a steel wire, akin to a paper lantern. After the lesson, the globe can be flattened for easy transport home.

194

In 1909, architect Bernard J.S. Cahill in San Francisco, made this radical attempt to create a globe that could be folded from one single piece of flat cardboard, the so-called Butterfly Map. It consists of eight rounded-off triangular maps. When folded together, it forms an octahedral volume that approximates a sphere. The advantage is the simplicity of production, but also the fact that the unfolded map is less distorted than the traditional Mercator world map. Cahill patented his folded cardboard globe in 1913. Another patented version of the Butterfly Map was a print on flat rubber, which could be folded into a ball, a model that was never taken into production.

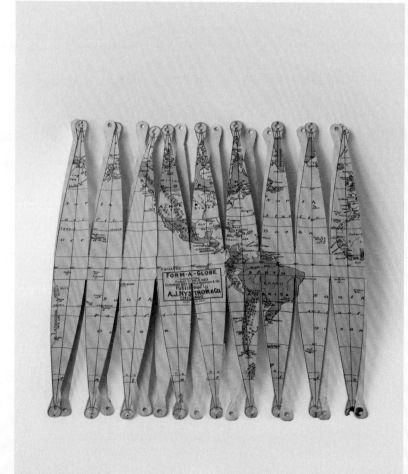

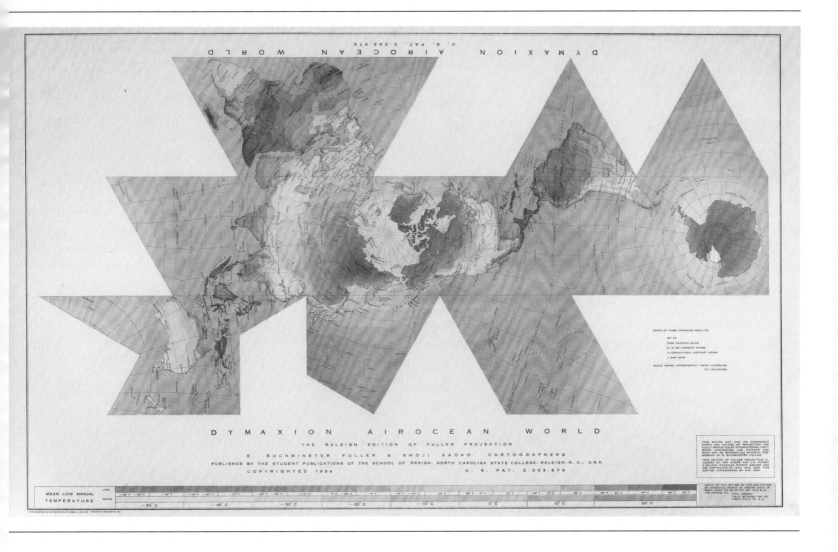

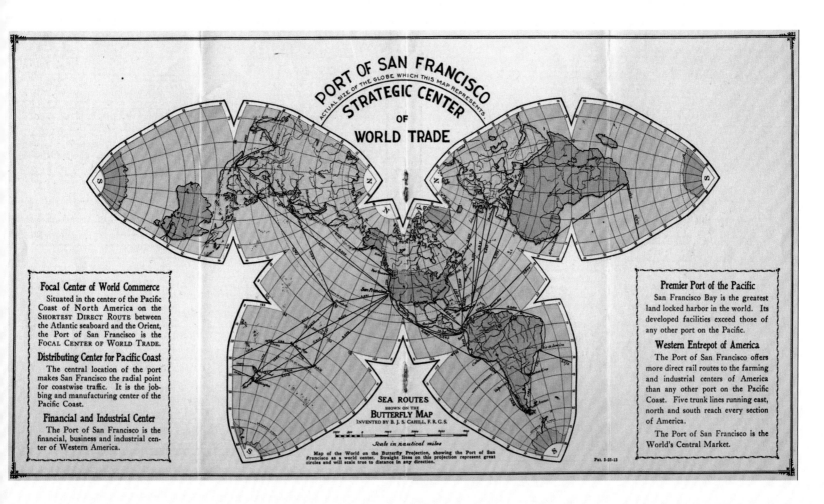

More Future Globes

195

British artist Luke Jerram created a seven-meter-diameter photographic globe called *Gaia*, which has been touring to various places around the world since 2018. The suspended globe rotates slowly, and its image is based on a photograph taken by Apollo 17 in 1972. It provides the public with the opportunity to experience the Earth as seen by an astronaut.

196

MOVA, a young Californian globe-making company, developed a four-and-a-half-inch self-revolving globe in the 2000s. This globe continuously rotates on a Plexiglas tripod, powered by sunlight. The globe features a stationary transparent acrylic outer scale with a rotating sphere within. The map image is a satellite photograph of the natural Earth.

197

Replogle created this custom-made Earth Geological Globe for the Commission for the Geological Map of the World in 2018. It showcases the most recent state of geological knowledge about the planet, based on computer data gathered by many international institutes. The colored surfaces represent geological time zones.

198

German artist Ingo Günther has been working on an ongoing project called *World Processor* since 1988. His aim is to visualize the global condition by using scientific data to create more than a thousand different illuminated thematic globes. Subjects range from political conflicts and social issues to the natural world and environment.

199

This twenty-first-century globe made in China levitates by means of a magnetic field. Early levitation globes had stands encasing the sphere. In this fifteen-centimeter model, the electromagnet is integrated into the base only. As long as it is powered by electricity, the small plastic sphere remains suspended in space.

200

This twenty-first-century lunar levitation globe made in China, has an electromagnet built in its faux wood base. it not only powers the twelve-centimeter Moon to float above the base but it also illuminates the 3D printed plastic relief sphere from inside, creating a convincing illusion of the Moon with its craters, radiating light in the dark.

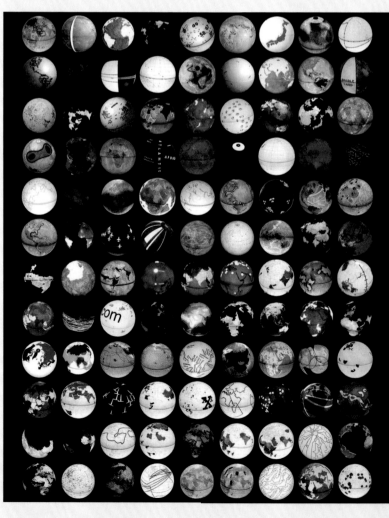

CREDITS

Cover image:
Norman Govoni climbs a ladder
to check on the condition of the Babson
Globe. Built in 1955 at Babson College
in Wellesley, Massachusetts, it was the
world's largest rotating globe at the time,
measuring 28 feet in diameter. /
Wellesley, 1970s

Dimensions of globes
The dimensions in this book refer to the diameter
of the spheres, without additional elements such as
meridians or stands. Twentieth-century machine-
made globes were mostly produced following
standardized sizes. For that reason, the size of
American globes is stated in inches and the size of
European globes is stated in centimeters. As an aid,
here is a conversion table with rounded dimensions:

Inches	Centimeters
4'	10 cm
6'	15 cm
8'	20 cm
10'	25 cm
12'	30 cm
14'	35 cm
16'	40 cm

Dating of globes
In the twentieth century, identical globes were
produced in large numbers over a period of time,
spanning several years. For commercial reasons,
globe makers rarely printed the production date on
their globes. As a consequence, it is impossible to
determine the exact production year of every globe
that is shown in this book. For that reason, the globes
are dated by decades, indicating the period in which
a specific model was sold and produced.

Origin of globes
All globes in this book are part of the collection
of Willem Jan Neutelings, except nos. 99, 113, 141, 181,
185, 190, 191, 192, 194, 195, 198.

AROUND
THE WORLD
IN

200
GLOBES

The author wishes to thank
Terenja van Dijk, Jan Mokre and Jaap van Triest
for their advice and support.

Research, composing and writing
Willem Jan Neutelings

Final editing
Michael Meert
Faina Peeters

Photography
Anne Deknock

Graphic design
Oeyen & Winters

D/2024/12.005/14
ISBN 9789460583674
NUR 680

Image from catalog of A.J. Nystrom & Co. /
Chicago, 1916
"A severe test on a Johnston Globe.
One of the reasons why 'Johnston's'
are worth more."

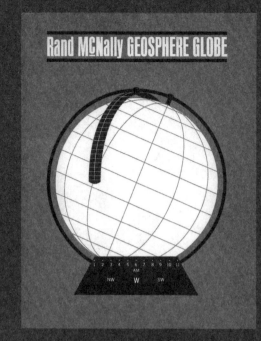